T0301702

PROCESS SAFETY MANAGEMENT

LEVERAGING NETWORKS AND COMMUNITIES OF PRACTICE FOR CONTINUOUS IMPROVEMENT

PROCESS SAFETY MANAGEMENT

LEVERAGING NETWORKS AND COMMUNITIES OF PRACTICE FOR CONTINUOUS IMPROVEMENT

Dr. Chitram Lutchman
Douglas Evans
Dr. Rohanie Maharaj
Ramesh Sharma

CRC Press is an imprint of the
Taylor & Francis Group, an **informa** business

CRC Press
Taylor & Francis Group
6000 Broken Sound Parkway NW, Suite 300
Boca Raton, FL 33487-2742

Printed on acid-free paper
Version Date: 20130624

International Standard Book Number-13: 978-1-4665-5361-3 (Hardback)

Library of Congress Cataloging-in-Publication Data

Lutchman, Chitram.
Process safety management : leveraging networks and communities of practice for continuous improvement / authors, Chitram Lutchman, Douglas Evans, Rohanie Maharaj, Ramesh Sharma.
pages cm
Includes bibliographical references and index.
ISBN 978-1-4665-5361-3 (hardcover : alk. paper) 1. Industrial safety. 2. Continuous improvement process. 3. Business networks. I. Evans, Douglas. II. Maharaj, Rohanie. III. Sharma, Ramesh. IV. Title.

T55.L878 2013
363.11--dc23 2013018221

Visit the Taylor & Francis Web site at
http://www.taylorandfrancis.com

and the CRC Press Web site at
http://www.crcpress.com

This book is dedicated to all workers who work tirelessly on a daily basis to improve business and health and safety performance. The book is especially dedicated to health and safety professionals and technical subject matter experts who devote much of their lives to keeping us safe both at work and during our private lives, and finding ways of doing things better. This book is also dedicated to the many workers who have perished in tragic process safety incidents, and to the families and friends who they have left behind.

Contents

Preface

Today's complex and ever-changing business environment requires organizations to continually evolve to meet business needs and to remain competitive. Failure to do so can ultimately lead to a slow and certain demise. Furthermore, stakeholders are sensitive and intolerant to business, Process Safety Management (PSM), and Management System inefficiencies. They are even more intolerant to preventable and avoidable catastrophic incidents. Networks and Communities of Practice within organizations are perhaps among the most underutilized opportunities for generating continuous improvements in business processes and practices. The authors provide a road map for organizations to identify and set up critical Networks for generating continuous improvements, preventing catastrophic incidents, and for sharing knowledge in an organized manner within the organization to enhance business performance. The information provided in this book is intended to help organizations establish centers of excellence by activating Networks for generating best practices and practical solutions to workplace business and safety challenges. The knowledge provided in this book covers the full range of activation of Networks including identifying members, defining goals and objectives, and prioritizing work through leadership and stewardship of Networks. The authors leverage practical and proven practices that are flexible and accommodating to varying scales of operations and industries. The expertise provided in this book is valuable to business leaders, engineers, HSSE (health, safety, security, and environment) practitioners, and students in all areas of engineering, business management, and health and safety. The ultimate goal of the authors is to help businesses along the path to more quickly address the workplace challenge of reducing workplace personnel and PSM incidents, of which 90% to 95% are predictable and avoidable, and 80% to 85% are repeated.

Authors

Chitram "Chit" Lutchman (principal author), DBA, MBA, Certified Safety Specialist (CSP), Canadian Registered Safety Professional (CRSP), 1st Class Power Engineer, B.Sc., is a 2011 recipient of the prestigious McMaster Arch Award for innovative accomplishments and contributions to society in Canada. Dr. Lutchman is an experienced safety professional with extensive frontline and leadership experience in the energy industry. Having sustained some disability from an industrial work accident, he developed a passion for improving health and safety in the workplace. With international oil and gas knowledge, Dr. Lutchman has experienced the two extremes of organizational health and safety practices, and business performance. As an employee of Canada's largest oil and gas producer, he pioneered work within his organization aimed at improving contractor health and safety management. In prior roles, he has functioned as a corporate leader in loss management and emergency response. He also functioned in project management leadership roles in the commissioning and startup of Canada's first commercial steam-assisted gravity drainage (SAGD) and cogeneration facility in Fort McMurray, Alberta. Dr. Lutchman published his first book, *Project Execution: A Practical Approach to Industrial and Commercial Project Management* (Taylor & Francis), in 2010. In 2012, he was the principal author of another book, *Safety Management: A Comprehensive Approach to Developing a Sustainable System* (Taylor & Francis). He is also the principal author of this book and has coordinated the contributions of his coauthors to provide readers with this practical package of information that can contribute significantly to improving the health and safety of all workers including new graduates, safety professionals, and business leaders and contractors. Dr. Lutchman is passionate about sharing knowledge in an organized way in the areas of HSSE, PSM, and Management Systems to ensure the health and safety of our most precious assets—people (in particular Generation Y)—while generating superior business performance for all stakeholders.

Rohanie Maharaj, Ph.D., M.Sc., M. Phil., B.Sc., is a graduate of the University of the West Indies (UWI) with degrees in natural sciences (B.Sc. Hons.) and food technology (M.Sc. and M. Phil.). She holds a Ph.D. in food science and technology specializing in food processing/postharvest technology from the Université Laval in Québec, Canada. She is a senior member of the American Society for Quality (ASQ) and the Institute of Food Technologists (IFT). Dr. Maharaj is also trained in Six Sigma methodology. At the University of Trinidad and Tobago (UTT), she has been employed as an associate professor since 2009 and is currently the program leader for food science and technology in the Biosciences, Agriculture and Food Technologies

Unit. Dr. Maharaj lectures in the areas of quality assurance, environment, health, and safety in the Petroleum Engineering Unit. She has more than 12 years of industry experience where she has worked in the manufacturing sector leading quality and reliability. Prior to joining the University of Trinidad and Tobago, she was a regional director and member of the board of Johnson & Johnson Caribbean. She has extensive industrial experience working in various capacities such as operations director, process excellence director, compliance/regulatory director, and quality assurance manager at Johnson & Johnson facilities in Trinidad, Jamaica, the Dominican Republic, and Puerto Rico. Dr. Maharaj received the Award of Excellence in 2005 at Johnson & Johnson for achieving exceptional business results. She was also instrumental in obtaining the ISO 14001 environmental certification for the Trinidad manufacturing facility in 2004. Apart from the many environmental and health and safety trainings offered at Johnson & Johnson, she has completed a number of international trainings including Behind the Wheel (BTW), International Health Care Compliance, Process Excellence Examiner, Marketing Companies Compliance, Green Belt, and Crisis Management. Dr. Maharaj currently serves on the Academic Council of UTT and is a member of the Food and Beverage Industry Development Committee (FBIDC) of Trinidad and Tobago.

Douglas J. Evans, P.Eng., B.Sc., is a graduate of Dalhousie University and the Technical University of Nova Scotia with B.Sc. and chemical engineering degrees in 1974 and 1976, respectively. He established his working career in the oil industry with Texaco Canada in Montreal, Quebec. During his employment with Texaco, Imperial Oil, Petro-Canada, and now Suncor, he has had the opportunity to experience Canada's cultural and geographic variety, with multiple work assignments in Eastern and Western Canada. Evans has held a variety of senior leadership roles in the areas of engineering, technology, strategic planning, supply, and operations management over his 37-year career. He was appointed director of technical services in Suncor's Calgary corporate offices in 2009 after the merger with Petro-Canada. Most recently, he was promoted to General Manager of Technical Services for Suncor. This role involves providing a center of excellence in process engineering and operational controls to support safe, reliable, and low-cost operations at all of Suncor's operating facilities, and for enhanced business performance. Evans's current focus is leveraging Suncor's technical excellence Networks to capture and share knowledge and best practices in order to reduce risk, build competency, and accelerate improvements across Suncor's business. Evans also has responsibility for stewarding process safety standards compliance and risk reduction across Suncor. His passion for Process Safety Management and risk reduction was also well utilized to drive deep organizational learnings across Petro-Canada as a result of the BP Texas City tragedy. His previous roles have included being director of process technology and reliability for Petro-Canada's Downstream Refinery operations, director of business integration for Petro-Canada Downstream, operations director at the Petro-Canada Oakville Refinery, area team manager of Crude Units and Powerformer at Imperial's Strathcona Refinery, and technical services manager at Strathcona Refinery. He has managed large expense and capital budgets, and has been responsible for safe and reliable operation of major refinery assets. What Evans enjoys most is working with talented and motivated people to achieve personal and business success. His reward is in supporting others to succeed.

Ramesh Sharma, BA, National Construction Safety Officer (NCSO), Canadian Registered Safety Professional (CRSP), Certified Safety Specialist (CSP), is a highly competent safety professional with demonstrated abilities in personnel and Process Safety Management. With more than 30 years experience in the field, Sharma has seen many predictable and avoidable work-related incidents. Having worked in mining and the oil and gas industry for most of his career, Sharma has a strong desire to contribute to improving organizational safety and business performance. He is a trained and certified safety professional with certification in CSP, CRSP, NCSO, British Columbia Institute of Technology (BCIT) Safety Certification, and Paramedic Level 1. He also holds instructor certification in various first aid and safety courses. A strong and passionate communicator, Sharma has developed, led, and implemented Safety Management Systems and programs in the mining industry as well as across several business areas in the oil and gas industry. A strong advocate for continuous improvements in business practices, business standards, and operating procedures, Sharma strives tirelessly to learn from every incident, aggressively investigating each incident so that lessons can be shared in an organized manner to prevent a repeat of similar incidents. In his more recent career, Sharma played a key role in setting up Process Safety Networks for Canada's largest oil and gas company—Suncor Energy. In addition, he functions as the Core Team Leaders for the Management of Change and Pre-Startup Safety Review Networks in Suncor. He has devoted much of his working life to creating a safer work environment for all workers and contractors. Sharma's motto: Good safety leads to great business performance. We must be diligent about preventing all incidents; working together, we can create a safer workplace for all. When incidents occur, we must seek to understand why they occurred, learn from them, and proactively put measures in place to prevent recurrence.

Acknowledgments

Formal organizational Networks for generating continuous improvements in business processes and practices are a relatively new concept in today's business environment. In view of this, the authors would like to acknowledge the contributions of their many peers and colleagues who have knowingly and inadvertently influenced their knowledge and expertise to this book. Among such pioneering contributors are members of DuPont Canada, Suncor Energy Inc., and other independent consultants. To these contributors, the authors are grateful and express their sincere thanks. The authors would also like to acknowledge the editorial contributions of Kevan, Megan, and Alexa Lutchman, who each took turns at grammatical and readability reviews. Their contributions helped to enhance the simplicity of this work that is now presented in a manner suitable to all audiences. Special thanks are also extended to Nishal Sankat for his online research contributions.

Introduction

Lutchman, Maharaj, and Ghanem (2012) have pointed out that 90% to 95% of incidents are avoidable and 80% to 85% are repeated. These incidents generally originate from two major sources, namely, *unsafe behaviors and actions* and *unsafe conditions*. Furthermore, data suggests the direct cost of incidents in the United States ranges from $50 to $60 billion per year with annual full cost estimates in the range of $180 to $200 billion (Lutchman et al., 2012).

The continued prevalence of global major incidents (most recently the 2010 BP Gulf of Mexico oil spill) and the preponderance of workplace fatalities and injuries, globally, begs the question: Why do incidents continue to occur in today's technologically advanced era? More important, with 80% to 85% of incidents being repeated, the more obvious questions are as follows:

1. Why do organizations fail to learn from prior incidents internal to the business?
2. Why do organizations fail to learn from their peers and other same-industry players?
3. Why do organizations fail to learn from the incidents and experiences of other industries?

The answers to these questions provide opportunities to generate tremendous improvements to business performance and to create a learning culture within organizations.

In this book, the authors draw upon field experience and knowledge to introduce a new and growing area of opportunity for organizations to improve upon health and safety performance, and overall business performance. The introduction of Networks and Communities of Practice for generating and disseminating knowledge within the organization is a proactive way to prevent incidents from occurring within the organization, while at the same time generating superior business performance. These Networks add value in the continual improvement of business performance in the following ways:

1. Learning from incidents that occur *within the organization*, and developing practical and financially viable solutions to prevent a repeat of the same or similar incident within your organization.
2. Learning from incidents that occur in organizations *within the same industry*, and applying practical and financially viable solutions to prevent the same or similar incidents from occurring within your organization.

3. Learning from incidents that occur *in other industries that can likely occur in your business or organization*, and applying practical and financially viable solutions to prevent the same or similar incident from occurring within your organization.
4. From existing operations, *finding new ways and means to improve* the reliability, operability, and performance of existing assets, technologies, and processes.

Networks, though relatively new to many organizations, have been in existence for decades. However, they exist as informal, unstructured, and underutilized processes known and used by few people within the organization. Furthermore, the full range of Network opportunities has not been leveraged within organizations, and this brilliant learning and value maximizing opportunity continues to be missed by many organizations and industries.

How Do We Define Networks?

Since the earliest of times, informal Networks have existed as a means of generating improvements and social well-being. It is difficult to imagine the life of early human beings on this planet without Networking (i.e., sharing knowledge, trusting each other, working in a cohesive way to find better solutions and improve human welfare). Moreover, informal Networks have existed in the workplace as a means of achieving business objectives. Today, Networking is an absolute requirement in our highly competitive environment.

We all maintain our own informal Networks with which we work every day. For example, if we are planning to buy a car or any other major expense item we will connect with someone *whom we trust* and from whose *advice we have benefited from* in the past. Also, members of the Network may be *more knowledgeable* (subject matter expert) regarding that product or service. In the purchase of a car, your informal Network may include some of the following members:

- Other users and stakeholders
- Financial experts for the best financing options
- Subject matter expert—a trusted individual
- Key experts in the dealership
- Friends who may have a similar vehicle to provide feedback on reliability, economy, and other desirable and undesirable attributes

Although these Network members may not all be consulted at the same time, can you imagine how quickly and effective the decision-making process would be should we have all members of this Network group at the same meeting to provide expert opinion on the decision-making subject?

Organizational Networks are intended to do just that. Networks are considered formal Networks when chartered with:

- A clearly defined scope of work
- Resourced with competent personnel, time, and money to achieve the goals
- Deliverables and properly stewarded

They harness the experience, knowledge, and subject matter expertise in an organized manner to create value for the organization.

Formal Networks in businesses are in the early stages of development within organizations. Where created, organizations have assessed the potential value creation from such units for improving processes and business practices to generate efficiencies, productivity, and overall performance.

Although informal and formal Networks may be characterized by different operational and organizational structures, it is important to point out that they both exhibit the following common attributes:

- Trusting relationships within and external to the Network
- Ability to influence the organization
- A Network leader with strong transformational leadership skills and capabilities
- Ability to select and prioritize among alternatives (scientific approach)
- Behaviors that leverage engagement, involvement, and collaboration
- Willingness to share the information
- Strong desires to improve
- Discipline
- Willingness to consider things from another Network member's viewpoint
- Ability to articulate and share the big picture
- Willingness to set and strive for aggressive goals

From a learning perspective, organizations have historically tried different approaches for addressing specific business issues in their quest for achieving best in class status and continuous learning. Among its people-related arsenal of tools were teamwork, committees, rapid response teams, and management Networks. Nevertheless, many of these were marginally successful and eventually faded away when leaders failed to recognize and derive immediate value.

History has also shown over time that Networks do work regardless of whether they are formal or informal. Organizations that have reached the world-class status have leveraged the concepts and attributes of formal Networks. Successful Networks were characterized by leadership commitment, operational discipline, clearly defined goals, and supporting resources. To enable successful Networks, they must be established with the right Network members, adequate Network structure, stewardship of Network activities, and conduits and tools for sharing of lessons with users throughout the organization.

The subsequent chapters in this book delve into the way Networks are created and provide the reader with a better understanding of how Networks are created and stewarded. The focus of this work is on formal Networks and it draws upon more than 100 years of collective experience of the authors. Networks provide a great opportunity for improving organizational performance in today's highly competitive, resource-constrained business environment.

Reference

Lutchman, C., Maharaj, R., and Ghanem, W. (2012). *Safety Management: A Comprehensive Approach to Developing a Sustainable System*. Boca Raton, FL: CRC Press/Taylor & Francis.

1

Process Safety Management (PSM)

The turn of the industrial age not only brought about increased demands for goods and services but a need to increase industries and technology. As a result, millions of people became employed and enjoyed a higher standard of living over the years, and companies also increased their profits. As industries increased the production of goods and services for consumers using hazardous materials, there was also an increase in the number of accidents, deaths, and injuries at the workplace.

Although industrialization has its benefits, every year millions of dollars are spent to compensate workers for injuries sustained on the job. Many of the injuries reported are a result of process safety issues. These may be unexpected releases of toxic, reactive, or flammable liquids and gases, which are highly hazardous chemicals that affect the conditions for safe and healthy workplaces. Incidents continue to occur in various industries that use highly hazardous chemicals that may be toxic, reactive, flammable, or explosive, or may exhibit a combination of these properties (Occupational Safety and Health Administration [OSHA], 2000).

Successful organizations see good Safety Management Systems (SMS) as good business. Such organizations commit significant amounts of resources toward creating a safer work environment in which there is zero tolerance for:

- Damage to the environment
- Harm to people
- Damage to facilities, equipment, and assets, and corporate image and goodwill

In spite of these efforts, we continue to experience major incidents that are costly both in terms of personnel impact, and domestic and global financial impact. Table 1.1 provides a historical summary of some of the world's greatest disasters related to Process Safety Management (PSM).

Today, the importance of safety and PSM is recognized by most organizations. Organizations respond with Safety Management Systems and Process Safety Management Systems as society continues to demand higher levels of safety performance.

TABLE 1.1
Major PSM Incidents with Severe Human, Economic, and Financial Consequences

		Consequence	
Incident	**Year**	**Fatalities/Injuries**	**Cost**
Mexico City	1984	Over 550 fatalities	$26 million
Bhopal Disaster	1984	Over 4,000 fatalities; 200,000 potentially affected	N/A
Chernobyl Nuclear Plant	1986	31 fatalities; over 200,000 future cancer deaths	N/A
Piper Alpha	1988	165 fatalities	
Pasadena, Texas	1989	Over 23 fatalities	$797 million
Pemex, Mexico	1996	Multiple fatalities	$253 million; $8 billion loss to economy
Petrobras, Brazil	2001	11 fatalities	$300 million
Toulouse, France	2001	29 fatalities; 20,000 homes damaged	N/A
Formosa Plastics, Illinois	2004	5 fatalities	N/A
Texas City, Texas	2005	15 fatalities	$50 million fines
Macondo, Gulf of Mexico oil spill	2010	11 fatalities	$4.7 billion fines; >$20 billion spent in cleanup cost
Venezuela, South America	2012	41 fatalities; 200 homes damaged	N/A

Source: From Patel, J., 2012, *62nd Canadian Society Chemical Engineering Conference Incorporating ISGA-3*, Vancouver, British Columbia, Canada, October 14–17.

This chapter discusses the following:

- Successful organizations that exercise best practices in Process Safety Management
- The types of programs used by these organizations to attain best in class performance
- How did they learn from their mistakes?
- How do they share best practices within the organization?

These discussions will better enable us to understand the requirements for successful execution of PSM within organizations and how its benefits can be improved by best practices in the specific industries.

Process Safety Management (PSM)

A process hazard is a major uncontrolled emission, fire, or explosion involving one or more highly hazardous chemicals that presents serious danger to employees in the workplace. PSM is the application of Management Systems to the identification, understanding, and control of process hazards to prevent process-related injuries and incidents. Patel (2012) defines PSM as the "application of Management Systems and controls (standards, procedures, programs, audits, evaluations) to a process in a way that *process hazards are identified, understood and controlled*" (p. 6). Patel further suggests that success in PSM requires "a sustained effort by leadership at all levels" (p. 6) of the organization, and for global organizations, a single global PSM Management System may be best.

PSM is a complete system approach for effective control of process hazards in every function of an organization. PSM is not limited to the process industries. It is fair to say the principles of PSM span across industries involved in research, development, and communication and information management technology. Perhaps not all of the PSM elements apply to these industries, but the requirements for hazard management continue to be the same.

Unexpected releases of toxic, reactive, or flammable liquids and gases in processes involving highly hazardous chemicals have been reported for many years in various industries that use chemicals with such properties. Regardless of the industry that uses these highly hazardous chemicals, there is potential for an accidental release any time the chemicals are not properly controlled, creating the possibility of disaster (OSHA, 2000). The phrase "Process Safety Management" came into widespread use after the adoption of OSHA Standard 29 CFR 1910.119 *Process Safety Management of Highly Hazardous Chemicals* and has been law since 1992 (Macza, 2008).

The principal concerns of PSM are as follows:

1. Process hazards—Concerns such as fires, explosions, toxic gases, and unintended releases, and how they affect the workers in their use, storage, and disposal.
2. Safety—Concerns how the company uses or implements safety regulations and any program to reduce incidents and injuries in the production process.
3. Management—Concerns all people who have some measure of control over the process, such as employee participation, operating procedure, and management of change.

Many companies develop their own structure to this holistic system that best suits their safety requirements, but in general, a PSM system can contain 14 parts or elements. These PSM elements may be grouped into three main themes: facilities, personnel, and technology, as illustrated in Figure 1.1. Although organizations may place a lot of emphasis on *personal safety*, this may conceal the fact that *process safety* measures are adequate for protecting

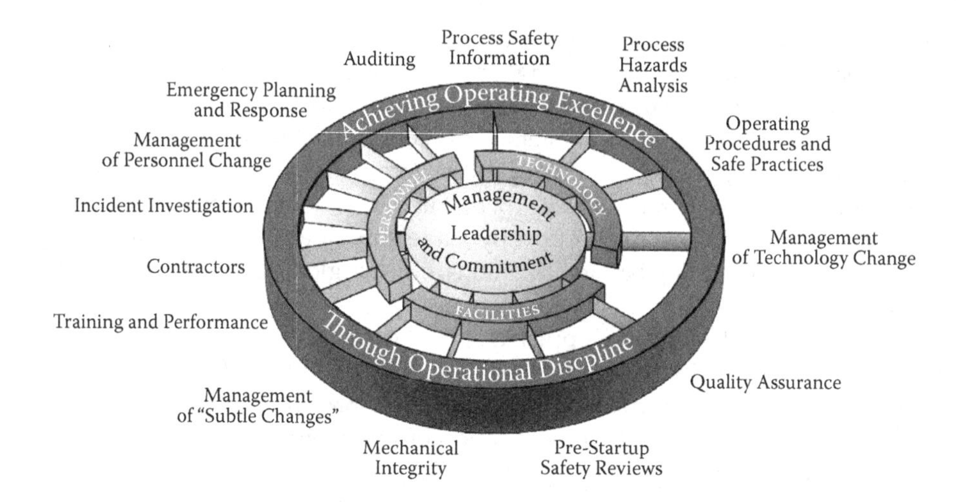

FIGURE 1.1
PSM elements categorized into facilities, personnel, and technology. (From Patel, J., 2012, *62nd Canadian Society Chemical Engineering Conference Incorporating ISGA-3,* Vancouver, British Columbia, Canada, October 14–17.)

people, the environment, and the company's assets and image once applied consistently and with the rigor required.

McBride and Collinson (2011) advised of a similar packaging of PSM elements into the categories of people, processes, and plant, as shown in Figure 1.2. They highlighted that this model was successfully applied for generating sustained high business performance at Centrica, an integrated energy company formed from the demerger of British Gas in 1997.

When major incidents are investigated, root cause analysis almost always points to a PSM failure or multiple PSM elements being compromised. In the 2005 Texas City incident, Turney (2008) advised, despite the fact that policies and comprehensive corporate procedures existed and there was a low personal accident rate, there was still a failure to address serious process safety issues that led to the tragic incident. One key element to PSM stability and functionality is a strong commitment from leadership and management, up to and including the board of directors being aware of and willing to act on safety concerns of the organization.

In order to address these issues, it is important that bold and courageous leadership prevails in the organization. Leaders must be prepared to forego short-term earnings to protect life, the environment, and assets. To do so they must have a good understanding of the principles in PSM, maintain effective and open lines of communication with all levels of the organization, and conduct continuous audits and assessment reviews predicated on the same principles of the plan–do–check–act quality improvement circle.

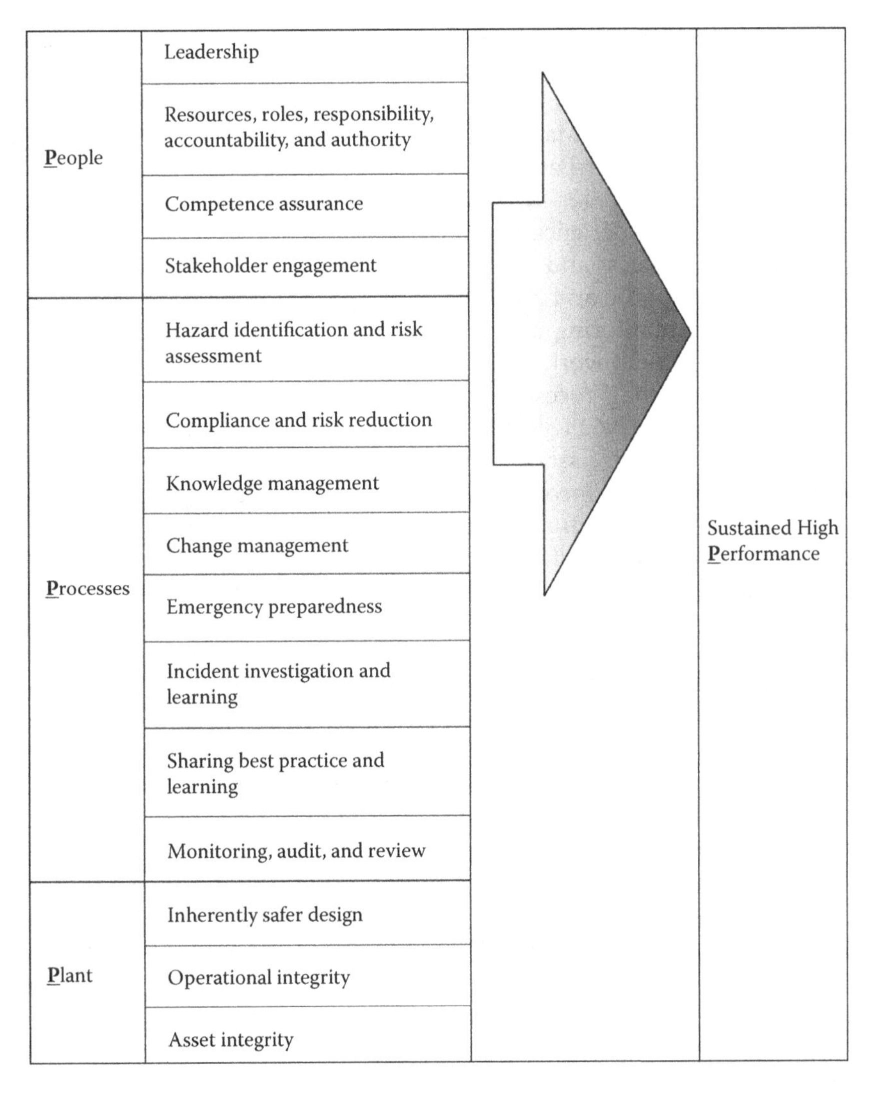

FIGURE 1.2
Centrica's approach to achieving excellence in process safety. (From McBride, M., and Collinson, G., 2011, Governance of process safety within a global energy company, *Loss Prevention Bulletin*, 217, 15–25. Retrieved November 3, 2012, from EBSCOhost database. With permission.)

Standards

Many organizations employ international and national safety standards as guidance to develop organization-specific Safety Management Systems (Wilkinson and Dale, 1998). According to the International Organization for Standardization (ISO), a standard is defined as "a documented agreement containing technical specifications or other precise criteria to be used consistently as rules guidelines, or definitions of characteristics to ensure that materials, products, processes and services are fit for their purpose" (Macza, 2008). The author also distinguishes *performance-based standards,* which are currently adopted in Canada and focus on what must be done, rather than on how it should be done. Performance-based standards are predominantly consultative, whereas *results-based standards* are traditionally more prescriptive standards as is experienced in the United States and Europe, which sets out details of the process and may or may not achieve the desired results.

According to Osborne and Zairi (1997), an SMS is composed of standards, procedures, and monitoring arrangements that aim to promote the health and safety of people at work. To help ensure safe and healthy workplaces, OSHA has issued the Process Safety Management of Highly Hazardous Chemicals standard (29CFR1910.119), which contains requirements for the management of hazards associated with processes using highly hazardous chemicals. It establishes a comprehensive management program that integrates technologies, procedures, and management practices (OSHA, 2000).

It is imperative that manufacturers involved in processes using chemicals, toxic gases, fluids, and other hazardous elements adhere to OSHA requirements in order to save lives, and keep their businesses compliant and in operation. OSHA's Hazard Communication Standard in 1983 gave the workers the *right to know* about workplace hazards, but the new Globally Harmonized System (GHS) gives workers the *right to understand* the workplace hazards. According to a March 2012 news release from the U.S. Department of Labor, the Hazard Communication Standard is being revised to align with the United Nations' Globally Harmonized System of Classification and Labeling of Chemicals, and will be fully implemented in 2016, further improving safety and health protections for America's workers. It will benefit workers by reducing confusion about chemical hazards in the workplace and facilitating safety training and improving understanding of hazards, especially for low literacy workers.

The GHS is also expected to prevent injuries and illnesses, save lives, and improve trade conditions for chemical manufacturers. OSHA's standard will classify chemicals according to their health and physical hazards, and establish consistent labels and safety data sheets for all chemicals made in the United States and imported from abroad. The revised standard also is expected to prevent an estimated 585 injuries and illnesses annually. It will reduce trade barriers and result in estimated annualized benefits in productivity improvements for American businesses that regularly handle, store, and use hazardous chemicals, as well as cost savings of $32.2 million for American businesses that periodically update safety data sheets and labels for chemicals covered under the standard (U.S. Department of Labor, 2012).

The implications for employers are as follows:

- Become informed about the standards and their updates
- Ensure employees are also informed and follow expected procedures

- Know the hazard classifications
- Know classification of mixtures
- Ensure that chemicals are properly labeled
- Use the safety data sheets
- Ensure that the workers be trained by December 1, 2013 (new label elements and safety data sheets)

Manufacturers of chemicals are still required to evaluate the chemicals they produce or import, and provide hazard information to the employer and employees. They are expected to put labels on containers and prepare safety sheets. Also, the modified standards set criteria for classifying chemicals according to their health and physical hazards. OSHA also requires a process hazard analysis of which the employer shall perform an initial process hazard analysis (hazard evaluation) on processes covered by this standard.

The process hazard analysis shall be appropriate to the complexity of the process and shall identify, evaluate, and control the hazards involved in the process. Employers shall determine and document the priority order for conducting process hazard analyses based on a rationale, which includes such considerations as the extent of the process hazards, number of potentially affected employees, age of the process, and operating history of the process. Since all accidents result from human error and errors by creating deficiencies in the design, the OSHA standard attempts to address the human factors in training and operating procedures. By the end of 2015, many employers around the world will be readily informed and be subjected to the penalties under OSHA.

PSM in Canada

Unlike the United States, where PSM is a regulatory requirement, Canadian PSM standards are derived primarily from the Canadian Chemical Producers Association's Responsible Care® program, which was developed in the 1980s. Responsible Care is a world-leading voluntary industry initiative. It is run in 53 countries whose combined chemical industries account for nearly 90% of global chemicals production. Every national Responsible Care program has eight fundamental features as outlined by Macza (2008):

1. A formal commitment by each company to a set of guiding principles—signed, in most cases, by the chief executive officer.
2. A series of codes, guidance notes, and checklists to help companies fulfill their commitment.
3. The development of indicators against which improvements in performance can be measured.

4. Open communication on health, safety, and environmental matters with interested parties, both inside and outside the industry.
5. Opportunities for companies to share views and exchange experiences on implementing Responsible Care.
6. Consideration of how best to encourage all member companies to commit themselves to, and participate in, Responsible Care.
7. A title and logo that clearly identify national programs as being consistent with, and part of, the Responsible Care concept.
8. Procedures for verifying that member companies have implemented the measurable or practical elements of Responsible Care.

Responsible Care can be used to bring a strong focus on SMS, particularly through planning, auditing, and reviewing alongside good performance reporting. This should assure company boards that the right things are actually being done at the plant in the right way. Below board level, high-quality training programs and staff involvement in hazard studies and process risk assessments will raise awareness of and control over the major hazards that employees are very close to every day of their working lives.

The Responsible Care program also encourages engagement with other industry stakeholders and in particular seeking opportunities for cooperation with regulators. Responsible Care needs to continue to lead the way into the future to promote the role of leadership within chemical businesses and, in particular, the importance of a positive safety culture, shared knowledge, and learning from cross-sector experiences (Chemical Industries Association, 2008).

The benefits of PSM in the United States have started to influence organizations in Canada. Although PSM is not at present a regulatory requirement in Canada, many organizations have voluntarily proceeded along the route of developing and executing PSM standards across their businesses. For these organizations, the message of good safety means great business appears to have infiltrated the organization with positive consequences.

Elements of PSM

According to OSHA (2010), a PSM program must be systematic and holistic in its approach to managing process hazards and must consider the following:

1. Process design
2. Process technology
3. Process changes

4. Operational and maintenance activities and procedures
5. Nonroutine activities and procedures
6. Emergency preparedness plans and procedures
7. Training programs
8. Other elements that affect the process

OSHA (2010) identifies the following PSM elements as required to effectively manage PMS risk exposures to an organization.

1. Employee Participation
2. Process Safety Information (PSI)
3. Process Hazard Analysis (PHA)
4. Operating Procedures
5. Training
6. Contractor Safety
7. Pre-Startup Safety Review (PSSR)
8. Mechanical Integrity
9. Hot Work Program
10. Management of Change (MOC)
11. Incident Investigation
12. Emergency Planning and Response
13. Compliance Audits
14. Trade Secrets

OSHA (2010) further advises that should small organizations address the following in their SMS, there is a greater chance that they will be compliant with regulatory requirements.

1. Process Safety Information (PSI)
   a. Hazards of the chemicals used in the processes
   b. Technology applied in the process
   c. Equipment involved in the process
   d. Employee involvement
2. Process Hazard Analysis (PHA)
3. Operating Procedures
4. Employee Training and Competency
5. Contractor Safety Management
6. Pre-Startup Safety Review

7. Mechanical Integrity of Equipment
    a. Process defenses
    b. Written procedures
    c. Inspection and testing
    d. Quality assurance
8. Nonroutine Work Authorizations
9. Management of Change (MOC)
    a. MOC—Engineered and nonengineered changes
    b. MOC—People
10. Incident Investigation
11. Emergency Preparedness Planning and Management
12. Compliance Audits
    a. Planning
    b. Staffing
    c. Conducting the audit
    d. Evaluation and corrective actions

Although OSHA does not differentiate into the categories defined in this section, it is easy to allocate PSM elements defined by OSHA into people, processes and systems, and facilities and technology. Packaged somewhat differently, compliance to PSM regulations can be achieved as shown in Table 1.2, which is recommended by Lutchman, Maharaj, and Ghanem (2012) for world-class safety performance.

PSM: People Requirements

Fulfilling regulatory requirements keeps organizations compliant with the law. However, most organizations seek to exceed regulatory requirements in a genuine quest for optimizing the health and safety of workers with its SMS. When developing the SMS it is best to ensure that at the very least, regulatory requirements are met. Several elements of PSM are focused on personnel management strategies for ensuring the health and safety of all workers. As shown in Table 1.2, these include:

1. Employee training and competency
2. Contractor Safety Management
3. Incident investigations
4. Management of change (personnel)
5. Emergency preparedness, planning, and management

TABLE 1.2

OSHA PSM Requirements: People, Processes and Systems, and Facilities and Technology Requirements

OSHA PSM Requirements	People	Processes and Systems	Facilities and Technology

• Employee training and competency
• Contractor Safety Management
• Incident investigation
• Management of change (people)
• Emergency preparedness, planning, and management
• Management of change (engineered/nonengineered)
• Nonroutine work authorizations
• Pre-startup safety review
• Compliance audits
  • Planning
  • Staffing
  • Conducting the audit
  • Evaluation and corrective actions
• Process safety information
  • Hazards of the chemicals used in the processes
  • Technology applied in the process
  • Equipment involved in the process
  • Employee involvement
• Process hazard analysis
• Operating procedures
• Mechanical integrity of equipment
  • Process defenses
  • Written procedures
  • Inspection and testing
  • Quality assurance
	• Employee training and competency
	• Contractor Safety Management
	• Incident investigations
	• Management of change (personnel)
	• Emergency preparedness, planning, and management
		• Management of engineered change and nonengineered change
		• Nonroutine work authorization
		• Pre-startup safety reviews
		• Compliance audits
		• Planning
		• Staffing
		• Conducting the audit
		• Evaluation and corrective actions
			• Process safety information
			  • Hazards of the chemicals used in the processes
			  • Technology applied in the process
			  • Equipment involved in the process
			  • Employee involvement
			• Process hazard analysis
			• Mechanical integrity of equipment
			  • Process defenses
			  • Written procedures
			  • Inspection and testing
			  • Quality assurance
			• Operating procedures

Source: Lutchman, C., Maharaj, R., and Ghanem, W., 2012, *Safety Management: A Comprehensive Approach to Developing a Sustainable System*, Boca Raton, FL: CRC Press/Taylor & Francis.

Employee Training and Competency

OSHA (2010) requires all workers (employees and contractors) working with hazardous chemicals and processes to understand the health and safety hazards associated with their work. Essentially, workers must be competent to do the work they are expected to safely perform and be able to protect themselves, coworkers, and communities from unintended releases from the process systems. Training and competency must extend beyond immediate hazards of the process fluids to include "operating procedures and safe work practices, emergency evacuation and response, safety procedures, routine and nonroutine work authorization activities" (OSHA 2010, para. 35).

Training programs must identify roles that are to be trained, training goals and objectives, and pre-established competency assessment requirements. Competency is reflected in the worker's ability and demonstrated capability to accomplish assigned work consistent with procedural and safe work requirements. Success in training is achieved when workers are engaged and the process is interactive. Compliance in training and competency requires organizations to evaluate the training programs on some frequency to determine if the training objectives are being met. Where failure rates are high, the training program must be reassessed and training repeated to ensure adequate knowledge transfer to workers.

The training program must be continuous and must be updated when changes to the process or operating systems have been made. Personnel must also be retrained when changes are made so that they can be competent in their work, inclusive of responses to the new operating requirements. Training and competency is perhaps one of the best investments organizations will make in its SMS. A trained and competent worker means there is a reduced likelihood for errors and mistakes and therefore a safer operating system. Furthermore, a trained and competent worker is a more motivated worker who will generally excel in workplace performance. Ultimately, training and competency creates a win–win situation for the organization regarding compliance and safety performance management.

In summary, from an organizational perspective:

- The objective of training and competency in PSM is to ensure that all workers responsible for operating and maintaining the operating systems are adequately trained and competent on how to safely do so.
- The training program should include provisions for initial and refresher training, and can provide an auditable process for verifying when training was done, the competency assessment process applied, and when follow-up training is due.
- Trained and competent personnel are an essential requirement for keeping process equipment and machinery operating safely.

- Initial training is intended to provide an overview and the competency required to operate safely with the known process, specific health and safety hazards using operating procedures (including emergency operation and shutdown), and safe work practices.
- Refresher training is intended to maintain the competency of workers over time at an appropriate frequency, generally not exceeding 3 years.

Contractor Safety Management

Where contractors are used in hazardous processes, they must be equally trained and competent as workers to ensure that they can protect the health and safety of themselves and other workers while performing assigned work. An effective contractor prequalification, screening, and verification process is required such that contractors with verified strong safety performance are used to perform work for the organization. Contractors must be similarly trained and competent as employees when performing hazardous work. Their training and competency records must be accessible and verified by organizations. Contractors' work must be authorized and controlled by the organization. In summary, from an organizational perspective:

- The objective Contractor Safety Management in PSM is to ensure that all contractors are aware of all known and potential hazards involved in the process and how their actions can affect the health and safety of all personnel working on that particular site as well as those in the surrounding communities.
- Contract language should hold contractors accountable for compliance with organizational safety requirements. Organizations are responsible for prequalifying potential contractors before selecting them and require timely corrective actions on Contractor Safety performance on an ongoing basis once selected.
- Contractors must follow safe work practices to ensure the health and safety of all workers and be compliant with regulatory requirements before work is done for the organization.
- Verify that contractors are trained and competent in the work they perform, and have access to all relevant procedures and information for completing work safely.

Incident Investigations

Incident investigation seeks to identify root causes of incidents or near misses such that preventative measures can be established to prevent recurrence. OSHA (2010) requires organizations to conduct investigations that have resulted in catastrophic events as well as those (near misses) that have the potential to do so. OSHA recommends the development of in-house capabilities for incident investigations by maintaining a multidisciplinary team of trained and competent incident investigators, such that timely incident investigations can be done. Investigations should be consultative and reports generated from investigations should be made available for Shared Learning. Investigations should also be focused on fact finding as opposed to blame determination. In summary, from an organizational perspective:

- The objective of the incident investigation is to provide learning and prevention opportunities such that similar incidents will not be repeated.
- Persistent investigation of all serious and near-miss incidents is essential for continuous improvements in the health and safety, and safety performance for the organization.
- Findings must be shared in an organized manner to promote learning and to identify and exploit gap closure opportunities where similar weaknesses in the SMS may exist across the organization.
- Incident investigation must be timely. A multidisciplinary team must be assembled with subject matter expertise comprised of relevant site personnel and nonsite personnel, including contractors if involved.

Management of Change (Personnel)

Management of change (personnel) (MOCP) considerations for personnel is generally an area overlooked by organizations that is often left to the human resources department to steward. MOCP is an important component of PSM since the availability of trained and competent personnel is required at all times for safe process operations. MOCP applies to critical roles in the following human resources management situations:

- When personnel are promoted or transferred
- When personnel resign or cease to work for the organization
- During situations of worker illnesses or deaths
- During downsizing and structural reorganization

Ultimately, the goals of MOCP are to ensure the availability of trained and competent personnel at all times to function in critical roles. Short-term solutions such as overtime are not sustainable and effective succession planning is essential to provide a workable solution for critical roles.

In summary, from an organizational perspective:

- The safe operations of any process facility require well-trained, competent, and experienced personnel.
- A balanced level of expertise is required at all times for safe operations of process facilities. Work groups in process facilities must at all times maintain a minimum level of collective experience and knowledge for the safe and continuous operations and maintenance of the facility.
- Loss of minimum levels of experience and knowledge through personnel movements and organizational changes increases the risk and hazards exposures of the process facility.
- Turnover management strategies are critical in MOCP and must be applied to critical roles in the process operations of the organization, such that the required minimum level of skills is maintained.

Emergency Preparedness, Planning, and Management

Emergency preparedness, planning, and management is essential in PSM since unplanned events will at some time occur with the release of energy, process fluids, and hazards from process systems. Organizations must be ready for such events to minimize the impacts. According to OSHA (2010), emergency response is the third line of defense activated when both the second line (control the release of chemical) and first line (operate and maintain the process and contain the chemicals) have failed.

Emergency management planning requires that all workers know what to do and how to respond in the event of an emergency, and they must be so trained. A warning device is required; a muster point or several muster points are required generally upwind of the potential hazard releases. Drills and mock emergency responses are required to develop the skills of those who are required to control and manage the response as well as those who are being evacuated. Personal protective equipment (PPE) must be available for all responders who will be exposed to the hazards and they too must be adequately trained and competent in the use of the PPE.

Where necessary, community involvement and mutual aid responders should participate in drills. Mutual aid responders are the collective response teams from other organizations who would assist in the event of an emergency in return for reciprocal emergency response support. Mutual aid responders are common in areas where multiple organizations are involved in hazardous businesses in the same region or within close proximity to each other. An incident commander and command post (or center) that is protected and outside the hazard zone is essential to support the response. Supporting tools such as worker telephone lists, external support resources list, emergency budget, and information technology (computers, radios, and telephones) must also be available in the command post. In summary, from an organizational perspective:

- Detailed emergency response planning for potential emergencies is required so that timely and effective response by the organization can protect workers, communities, the environment, and facilities, which may be exposed to hazards during an incident.
- Emergency response planning must consider the many possibilities of emergency situations and must be well prepared to respond to any and all of these possibilities.
- All responders must be properly trained and competent in their response roles. They must also have full and ready access to all PPE required to support the response and must be trained and competent in the use of PPE.
- Training exercises, simulations, and planned and unplanned emergency response drills are vital parts of emergency planning, which helps to develop the readiness of the response teams.

PSM: Processes and System Requirements

Success in PSM requires the involvement of many supporting systems and processes. This section introduces some of the key processes and systems that enable PSM. Processes and systems include the following as per Table 1.2:

1. Management of engineered change and nonengineered change
2. Nonroutine work authorization
3. Pre-startup safety reviews
4. Compliance audits

Management of Engineered Changes and Nonengineered Changes

According to OSHA (2010), when we consider PSM as changes, other than *replacement-in-kind*, we refer to the following:

- Modifications to equipment and machinery
- Changes to operating procedures
- Changes in the types of inputs and raw materials to the process
- Changes to operating and processing conditions

Management of change (MOC) is a systematic procedure that organizations adopt to ensure all hazards are addressed when change is made. Generally, when change is made in process organizations, it may involve changes in process, operating limits (temperatures, flows, or pressure), or equipment. When such changes are made, new hazards may be introduced into the process.

The MOC process is designed to identify these newly introduced hazards such that they may be adequately mitigated to reasonably practicable levels. Included in the MOC process is the assignment of action items and due dates for completion of mitigation actions before the facility or process is restarted after the change. In addition, where necessary, training and competency evaluation is required before the change can be activated. Failure to apply an effective MOC process in organizations is among the leading causes of catastrophic incidents in organizations.

Management of Engineered Change

Engineered change refers to any change that is a specified addition, alteration, or removal of equipment, facilities, infrastructure, or software. Alternatively, it refers to any change to standards or specifications that may result in new components, materials, processes, or procedures being introduced. This includes any alterations to the process safety information for a hazardous process. For most organizations, where engineered changes are concerned, the following prerequisites are generally addressed before changes are made:

- Purpose of change. Why is the change necessary? Is there a business case for doing so? The business case may include but is not limited to eliminating a process hazard.
- Assessment of risks associated with not making the change.
- Safety, occupational health, and environmental impact including whether a process hazard analysis is required. When a process hazard analysis is required, the completed and approved review must be attached to the management of change document.
- If the change involves a deviation from established standards, the technical deviation must be authorized and included with the management of change and the process safety information, amended accordingly.
- Modifications to operating procedures.
- Essential training and communication needed for employees involved in operating and maintaining the process and contract employees whose job tasks are affected by a change in the process shall be informed of and trained in the change prior to the startup of the affected part of the process.
- Limits for the change (time period and/or quantity of material).
- Approval and authorization requirements.
- Any applicable Process Safety Management regulatory coverage.

Trial runs or temporary changes that exceed safe operating limits shall require a management of engineered change. According to OSHA (2010), many temporary changes have become permanent changes and have resulted in catastrophic events. Temporary changes must, therefore, be carefully managed, and proper documentation, review, and evaluation considerations are essential before such changes become permanent, so that the health and safety of workers, the environment, and assets can be preserved. Processes should be returned to standard operating conditions following an authorized trial period. If permanent changes are recommended as a result of a trial run, a new management of engineered change must be initiated. Areas should establish and implement a tracking system to provide closure to those changes where there was either a trial or a test period. All engineered changes should require a pre-startup safety review (PSSR).

When properly applied the requirements of MOCs are guided by the following:

- The objective of MOC process is to ensure that all hazards introduced by the implementation of the change are identified and controlled prior to resuming operation.
- The MOC process is designed to ensure compatibility between process and equipment where changes are made.
- All changes to the equipment, process, and procedures are properly reviewed to ensure hazards introduced in the process can be and are adequately addressed.
- All personnel required to operate the facility or process after the change is made are properly trained and competent in operating the facility or process.

Nonengineered Changes

Nonengineered changes refer to changes introduced into the process where engineered specifications or process safety requirements are not affected. Nonengineered changes may include but are not limited to changes to control systems or operations, graphic configurations, equipment titles, descriptions, equipment, and similar related items. Nonengineered changes may also include small nonimpacting changes that are not *replacement-in-kind* to assets or software potentially impacting hazardous processes. Although we may consider nonengineered changes as relatively unimportant, they still constitute change; and over time, with additional changes new hazards and risks may be introduced into the process. As a consequence, all nonengineered changes must be documented in a similar manner as to engineered changes.

Ultimately, the bottom line in using an MOC process is that all modifications are subject to:

- Adequate review of the impacts of the change on the process and on process safety considerations
- All changes being appropriately documented

- All changes being authorized by appropriate levels of management
- All personnel being made aware of the changes and, where necessary, are appropriately trained and competent

As a consequence, at the very least, PSM elements affected must be considered when changes are made to include process safety information (PSI), pre-startup safety review (PSSR), and employee training and competency.

Nonroutine Work Authorization

What do we mean by nonroutine work in PSM? In process operations, there is routine work and nonroutine work. The daily work of plant and facilities operators is considered routine work and includes adjustments made by operations personnel to produce the desired final products. Nonroutine work, on the other hand, refers to all work that may be conducted periodically. Examples of nonroutine work may include work related to planned and unplanned maintenance, turnarounds, changes, and modifications to equipment and machinery.

When nonroutine work is conducted, such work must be properly monitored and managed to avoid incidents. Nonroutine work must be properly authorized in a consistent manner. A work permit that provides documentation of the scope of the work, hazards identification, and mitigation process are among the required guidance to whoever is performing the work is essential.

The work permitting process should include verification of procedures used for but not limited to the following:

- Isolation of any energy sources such as electricity, pressure, and temperatures
- Breaking the integrity of a closed system
- Confined space entry
- Hot work processes requirement and authorizations
- Work completion and return to service procedures

Nonroutine work authorization is perhaps one of the most important steps in maintaining the health and safety of personnel, environment, and assets during nonroutine work. *We must bear in mind that most incidents occur during instances of nonroutine work, and when incidents occur, work permits are among the first sets of documents that investigators will acquire to begin an investigation.* Care and attention must be provided to the work permitting process and due-diligence requirements are necessary for ensuring a complete and effective hazards assessment and mitigation exercise has been done, and that personnel are provided the right tools, procedures, and oversight when nonroutine work is performed. A well-documented, transparent process is required.

Pre-Startup Safety Review (PSSR)

Experience has shown that process safety incidents occur when plants and facilities are being shut down or when they are being started up. This can happen during Greenfield startups or shutdowns as well. To reduce the possibilities of such incidents, a facilities readiness review (Lutchman, 2010), or PSSR, helps in determining the readiness of the facilities to be started up. A PSSR is a jointly performed exercise between representation of multiple stakeholders and subject matter experts to verify the following prior to the startup and introduction of process hazards into new and modified process equipment:

- All new equipment and installation are completed in accordance with design specification.
- All required and affected safety, operating, maintenance, and emergency procedures have been updated and are adequate.
- All workers involved in the operation and maintenance of the new process, equipment, and/or facilities are properly trained and competency assessed.
- Modified facilities meet all PSM requirements. PSSR provides a final review of any new or modified equipment to verify all appropriate elements of PSM have been addressed satisfactorily and that the facility is safe to operate.
- PSSRs are generally conducted by multidisciplinary teams consisting of operations, technical, design, maintenance, and appropriate safety representatives on an as-needed basis.

A PSSR completion report must be accepted and approved by the senior facility leader prior to startup.

Compliance Audit

Ultimately, an audit is a process for gathering information that would guide business leaders on the degree of compliance or noncompliance to the PSM element standard and to all PSM standards. Compliance audits are proactive verification of compliance to PSM element standards. OSHA (2010) points to organizational requirements of an impartial, trained, and competent team of compliance auditors (or individual for smaller facilities) for performing comparative assessments between field actions and requirements of the accepted standard by which the organization is guided.

The aim of the audit is to proactively identify hazards that are undetected or ignored by businesses, and to take the necessary corrective actions to prioritize and address these hazards and risk exposures of the organization before incidents occur. Audits must be properly planned and done such that facilities being audited do not feel intimidated by the process. Leaders of facilities being audited must perceive audits as a positive process for improving his or her operations and overall risk and hazards exposures as opposed to a fault-finding process. As a consequence, engagement of personnel is encouraged and organizations must be forewarned and provided adequate opportunities to prepare for the audit. Well-run organizations with strong safety cultures welcome audits and see them as positive steps in improving the health and safety of the workplace.

When audits are conducted, emphasis should be on processes, transparency, and traceability for compliance to the standards. A check sheet for every standard helps in the auditing process so that all components of compliance to the standard can be reviewed. Audits include a series of interviews of personnel in leadership and frontline roles as well as workers at the frontline. Audits also include site inspection and physical inspection of the facilities. Interviews provide stakeholders understanding of the requirements of each standard and their related roles and responsibilities. It also provides the auditor opportunities to identify improvement opportunities and corrective actions.

Audits and compliance provide opportunities for identifying and prioritizing a list of compliance action items for stewardship by the business. Leadership review and debriefing of audit findings and action items is necessary, such that appropriate priorities, timelines, resource allocation, and responsibilities can be established. A historical database of action items is also essential since it provides opportunities to trend responses to compliance requirements and to preferentially allocate resources during stewardship if necessary. Traceability is essential to facilitate the audits and compliance process. In summary, from an organizational perspective:

- The objective of the auditing element of PSM is to evaluate the effectiveness of PSM by identifying deficiencies, risks, and hazards exposures of the organization and recommending corrective action. The audit also identifies pockets of excellence that can be leveraged across the entire organization.
- Audits provide a measurement of compliance with the established PSM standard for each element.
- An effective auditing approach requires a cultural shift for many organizations to conduct self-audits supported by audits performed by an independent impartial group of trained and competent auditors.
- Audits of each element of PSM should not exceed a finite period between audits. Typically, this may be 24 to 36 months. Care must be taken to ensure audit fatigue does not become a challenge that eventually evolves to a window dressing exercise to demonstrate compliance and fulfill an obligation.

PSM: Facilities and Technology

The safe and efficient operations of facilities and technology are also a core focus of PSM. As a consequence, standards have similarly been developed to ensure the safe and efficient operations of facilities and technology. In this section, we shall discuss these requirements and PSM element standards for facilities and technology. As shown in Table 1.2, these include the following:

1. Process safety information (PSI)
2. Process hazard analysis (PHA)
3. Mechanical integrity of equipment
4. Operating procedures

Process Safety Information (PSI)

PSI refers to the management process for ensuring the availability of complete and accurate written information concerning process chemicals, process technology, and process equipment. PSI is essential to an effective PSM process since it makes available all relevant information to all stakeholders involved in the stewardship, operations, and maintenance of a process facility or technology. PSI management includes the way information is managed regarding but not limited to the following:

1. Process flow diagrams (PFDs) and piping and instrumentation diagrams (P&IDs)—The diagrams are essential in ensuring processes remain within the operating parameters of the process and identifies the controls required for maintaining the process. Accurate and updated P&IDs, in particular, are essential for enabling timely containment and control in the event of an incident. Experience has shown also that P&IDs are generally very poorly updated when changes to the processes and facilities are made and in many instances several versions of P&IDs may be retained leading to major confusion and delays when trying to contain an incident.
2. Policies, codes of practices (COPs), standard operating procedures (SOPs), critical practices, and industry practices—These documents contain and provide information to workers about how work is to be conducted. Policies, COPs, and SOPs are critical documents for ensuring work is conducted safely at the facility. These documents must be easily accessible and available to all workers required to do routine and nonroutine work at the facility.
3. Materials safety data sheets (MSDS)—MSDS provide information on the hazards of all materials and chemicals used or processed at the worksite or facility. This information is essential to inform all workers of the hazards they are exposed to at the worksite.

4. Technology information—Technology information refers to the information relating to the technology used in the process. Technology information may include but is not limited to pre-established criteria for maximum inventory levels for process chemicals. Exceeding these limits may be considered abnormal operating conditions requiring a risk mitigating strategy.
5. Technical information on equipment, machinery, and interconnecting piping and devices—The focus here is on the codes and standards used to establish good engineering practices.

In summary, from an organizational perspective, PSI deals with:

- Compilation of written information that enables workers to identify and understand potential hazards they may be exposed to in the workplace
- PSI includes
  - Documentation of the related chemical hazards exposures and consequences
  - The technology used in the design and operations of the process
  - Documentation of the technical specifications of all equipment, machinery, and interconnecting pipelines
- Importance of PSI
  - PSI is foundational to PSM
  - Provides essential information for conducting PHA and stewardship of mitigation strategies
  - Critical for generating employee participation, compiling, reviewing, and updating operating procedures
- Critically important for demonstrating due diligence in audits and compliance

Process Hazards Analysis (PHA)

OSHA (2010) regards PHA as one of the most important elements of a PSM. PHA is an organized and systematic process for identifying and analyzing process hazards and developing mitigating strategies for reducing them to as low as reasonably practicable levels. PHAs are generally conducted by a multidisciplinary team of subject matter experts, who are led by an experienced facilitator such that all hazards and accompanying protection offered by the system are identified. Where engineering solutions cannot be included in the risk mitigation strategies, successive layers of protection that include administrative controls are developed before the last level of protection—PPE—is considered. The PHA team must be capable of putting aside personal goals and focus on finding the best solutions to hazards, designs, controls, and risk mitigation strategies. The team leader must also be skilled in consensus building such that true solutions can be developed for risk mitigation.

In summary, from an organizational perspective:

- A PHA team is multidisciplinary and comprised of subject matter experts in the process being reviewed.
- The PHA leader and facilitator must be experienced and knowledgeable in PHA methodologies and terminologies.
- The business leadership is ultimately accountable and responsible for determining when PHAs are required and for ensuring quality PHAs are conducted.
- A well-conducted PHA provides assurance that all process hazards are identified and adequately addressed through engineering and administrative controls in that order.
- The importance of PHAs cannot be understated.
    - PHAs identify potential hazardous outcomes and events, and require proactive solutions to these hazards.
    - PHAs provide opportunities for the best solutions for hazards—engineering controls.
    - PHAs help in identifying required actions necessary for eliminating and/or reducing process and operational risks in process facilities.
    - PHAs highlight the interrelationships among PSM elements for ensuring the health and safety of all personnel required to work on the facility.
    - PHAs provide transparency and due diligence to protecting the health and safety of workers, the environment, and facilities.

Mechanical Integrity (MI) of Equipment

The mechanical integrity (MI) of a process system is the outcome of the following:

1. Process defenses
    - Operating principles and guidelines—Essentially, the goal is to ensure the operating limits of temperatures, pressures, flows, and other specified operating limits and conditions are not exceeded.
    - Process built-in protection systems, such as pressure release systems, vents, overflow systems, and flares. In the event that operating principles and guidelines are exceeded, the built-in protective systems of the facility are activated to protect the facility.
2. Written procedures are intended to support the maintenance requirements of each category of equipment at the facility.

3. Inspection and testing requirements are determined by manufacturer specifications and applicable codes. Inspection and testing are proactive measures to examine the degree of wear and tear experience by equipment during fixed cycles of operation. Erosion and corrosion are common concerns for process equipment and machinery that must be assessed and evaluated to prevent failure. In many instances, inspection and testing frequency may exceed manufacturer recommendations.
4. Quality assurance helps to maintain the integrity of construction by ensuring the proper use of materials and techniques (such as welding). Quality assurance is essential in ensuring the defenses of the process.

According to OSHA (2010), key requirements of MI include:

1. Naming and categorizing equipment and instrumentation.
2. Establishing inspections and testing requirements and acceptance criteria for each piece of equipment and instrumentation, and the required inspection and testing frequency.
3. Development of required preventative and maintenance procedures.
4. A trained and competent maintenance workforce.
5. Adherence to manufacturer recommendations.
6. Documentation of test and inspection results.
7. Adherence to manufacturer recommendations for equipment and instrumentation.

In summary, from an organizational perspective:

- MI is essential to ensure the integrity of a process is maintained through adequate preventive maintenance and inspection and testing procedures to prevent unexpected equipment failures.
- MI requires demonstrated commitment to using written procedures, and for trained and competent personnel maintaining the ongoing integrity of the process equipment.
- MI includes a comprehensive approach to equipment inspection and testing, and documentation of findings such that appropriate frequency can be established. At the very minimum, manufacturer recommendations must be maintained.

Operating Procedures

Operating procedures or standard operating procedures (SOPs) are tools provided by the organization that clearly describe the sequential steps required for tasks to be performed such that all work can be done safely. SOPs also provide technically correct operating limits and normal operating conditions for each process that personnel must abide to. As part of the routine work, SOPs identify data to be collected, the frequency of data collection, and required samples to be collected. Ultimately, SOPs are provided to all workers to ensure that work is done safely at all times and the safe operation of any facility is maintained.

Through the involvement of those using these SOPs, they are revised periodically and always kept current through version controls. If an SOP has been identified to have an obvious error that can jeopardize the health and safety of workers, immediate steps must be taken to correct the deficiency in the SOP and to communicate the change to all users. When changes are required, the established MOC process must be followed. Where necessary, should training be required, competency assessment is also essential. SOPs must provide instructions on all safety requirements for performing each step of the task safely.

In summary, from an organizational perspective:

- SOPs are developed to ensure all nonroutine operating and maintenance work is conducted safely.
- SOPs are very effective training tools for developing the competency of inexperienced workers.
- SOPs must be made available and accessible to all workers (employees and contractors) at all times.
- When nonroutine work is being authorized, all required and relevant procedures for that work must be included in the work package and should be reviewed by the work group before any task is undertaken.
- SOPs must contain guidance about normal operation as well as required guidance about emergency operation.

In this chapter, we discussed the key elements of PSM, which deal with the following:

1. Preserving the health and safety of all workers at all times on a process facility.
2. The required processes and systems for ensuring all work is conducted safely at all times.
3. Maintaining the integrity of the facilities and processes such that the health and safety of all workers are maintained.

PSM introduces a bold and complete new approach for managing the health and safety of workers and for reducing the impact of capital losses associated with catastrophic failures and disasters in the workplace.

References

Chemical Industries Association. (2008). Process safety leadership in the chemicals industry. Chemical Industries Association, London. Retrieved August 23, 2012, from http://www.cefic.org/Documents/IndustrySupport/CIA%20Process%20Safety%20-%20Best%20Practice%20Guide.pdf.

Lutchman, C. (2010). *Project Execution: A Practical Approach to Industrial and Commercial Project Management.* Boca Raton, FL: CRC Press/Taylor & Francis.

Lutchman, C., Maharaj, R., and Ghanem, W. (2012). *Safety Management: A Comprehensive Approach to Developing a Sustainable System.* Boca Raton, FL: CRC Press/Taylor & Francis.

Macza, M. (2008). A Canadian perspective of the history of process safety management legislation. *8th International Symposium Programmable Electronic System in Safety-Related Applications.* Retrieved August 23, 2012, from http://www.acm.ab.ca/uploadedFiles/003_Resources/Articles_in_ProcessWest/A%20Canadian%20Perspective%20of%20the%20History%20of%20PSM%20Legislation.pdf.

McBride, M., and Collinson, G. (2011). Governance of process safety within a global energy company. *Loss Prevention Bulletin,* 217: 15–25. Retrieved November 3, 2012, from EBSCOhost database.

Occupational Safety and Health Administration (OSHA). (2000). Process Safety Management. U.S. Department of Labor, Occupational Safety and Health Administration. Retrieved August 23, 2012, from http://www.osha.gov/Publications/osha3132.pdf.

Occupational Safety and Health Administration (OSHA). (2010). Process safety management guidelines for compliance. Retrieved January 1, 2011, from http://www.osha.gov/Publications/osha3133.html.

Osborne, J., and Zairi, M. (1997). *Total Quality Management and the Management of Health and Safety.* Sudbury: HSE Books.

Patel, J. (2012). The effects of risk management principles integrated with a Process Safety Management (PSM) system. *62nd Canadian Society Chemical Engineering Conference Incorporating ISGA-3,* Vancouver, British Columbia, Canada, October 14–17.

Turney, R. (2008). Progress in process safety: 200 issues of LPB. *Loss Prevention Bulletin,* 200: 3–6.

U.S. Department of Labor. (2012). U.S. Department of Labor's OSHA revises Hazard Communication Standard: Regulation protects workers from dangerous chemicals, helps American businesses compete worldwide. Release Number: 12-280-NAT. Retrieved August 26, 2012, from http://www.osha.gov/pls/oshaweb/owadisp.show_document?p_table=NEWS_RELEASES&p_id=22038.

Wilkinson, G., and Dale, B. G. (1998). System integration: The views and activities of certification bodies. *The TQM Magazine,* 10(4): 288–292.

2

Historical Perspective: A Review of Operationally Disciplined and Excellent Organizations Where Process Safety Management Is Entrenched

Renowned author Jim Collins (2001), in his book *Good to Great*, emphasized the importance of discipline in his writing. Collins advised: "When you have disciplined people you don't need hierarchy; when you have disciplined thought you don't need bureaucracy; when you have disciplined action you don't need excess controls" (p. 13). It is no secret, therefore, that many organizations are today seeking to drive discipline in their business practices, processes, and Management Systems.

Operational discipline is the foundation of every successful business Management System and is considered the essence for achieving operational excellence. This chapter discusses the elements of operationally excellent and disciplined Management Systems that are endemic to several successful companies. World-class leading organizations, such as DuPont, Chevron, British Gas, and Exxon Mobil, are explored to better understand the value created from embracing discipline to generate operational excellence in business performance.

Operational Discipline

The concept of operational discipline was distilled from decades of observing the characteristics of high-performance teams and individuals. Operational discipline may be defined as a consistent pattern of behavioral choices that when demonstrated by leadership and supported within organizations leads to a cultural shift with accompanying organizational performance and success. It is the dedication and commitment by the organization to perform its work consistent with the requirements of the adopted Management System, and aligned and applied procedures. According to Klein and Vaughen (2008), discipline in this context does not refer to punishment but to doing things in a way that results in positive benefits.

Developing a high level of operational discipline is the essence for achieving operational excellence. Discipline is the foundation of every successful business management system. "When you combine a culture of discipline with an ethic of entrepreneurship, you get the magic alchemy of great performance" (Collins, 2001, p. 13). Product and service quality is naturally enhanced and profitability can be maximized when organizations consistently display the behaviors of operational discipline in its decision making and business practices.

According to Walter (2002), for many successful organizations, operations discipline is embedded in the way these organizations manage the following:

1. Economic risk management system
2. Sustainability programs
3. Behavior-based safety programs
4. Process Safety Management (PSM) programs
5. Environmental risk management

Other areas of operations discipline that generally contribute to discipline and excellence include:

1. Leadership and commitment
2. Stakeholder engagement and management
3. Change management
4. Personnel development; training and competency development

When operations discipline is embedded in the organizational *way of doing things* or *culture,* these organizations perform significantly better than their peers from profitability and goodwill viewpoints.

The authors regard PSM as a key component of operations discipline. Although all elements of PSM are not necessarily applicable to all businesses, the concepts of PSM are applicable to all businesses and the discipline derived from executing PSM across the business adds to the operational excellence of any organization. The next section provides a brief introduction into the history of PSM.

Process Safety Management

Process Safety Management Systems are only as effective as the day-to-day ability of the organization to rigorously execute system requirements correctly every time. The failure of just one person in completing a job task correctly just one time can unfortunately lead to serious injuries and potentially catastrophic incidents. In fact, the design, implementation, and daily execution of PSM systems are all dependent on workers at all levels in the organization doing their job tasks correctly every time.

High levels of operational discipline, therefore, help ensure strong PSM performance and overall operational excellence. Macza (2008) defines PSM as the applications of programs, procedures, audits, and evaluation to a manufacturing or industrial process in order to identify, understand, and control process-related hazard risks resulting in systematic improvements and safety standards. PSM is therefore a complete system approach for effective control of process hazards in every function of an organization. It is not limited to the process industries and can even include research and development, communication, and information management. Occupational Safety and Health Administration (OSHA) regulation 29 CFR 1910.119 (Process Safety Management of highly hazardous materials) was introduced in the United States as a legal requirement for process operations in 1992.

Many businesses develop their own structure to this holistic system to better suit their safety requirements. However, a PSM system can contain 14 parts or elements for achieving sustainable and consistent business success in process operations. Lutchman, Maharaj, and Ghanem (2012) grouped these PSM elements into three main categories: people, processes and systems, and facilities and technology (see Chapter 1, Table 1.2). The key element to PSM stability and functionality is a strong commitment from management and leadership.

According to Lutchman, Maharaj, and Ghanem (2012) and other proponents of PSM, process safety incidents can result in catastrophic failures and are generally associated with major process explosions, fires, and spills. PSM incidents are costly to both the organization and society. PSM incidents therefore must be avoided, and operations discipline helps to achieve this goal.

History of Process Safety Management

Sutton (2011) advised that PSM was borne out of the 1974 Flixborough disaster, a massive chemical plant explosion. The Flixborough plant, operated by Nypro since 1967, produced caprolactam, which was used to manufacture nylon. The process utilized six large pressurized reactors containing cyclohexane (a liquid in ambient conditions, which easily forms flammable vapor and releases significant energy when ignited). On March 27, 1974, a vertical crack appeared in one of the reactors and began leaking cyclohexane. An investigation revealed a serious problem with the reactor and a bypass was set up between two adjacent reactors to continue production. Engineers inexperienced in high-pressure pipework designed the bypass; no plans or calculations had been produced and the pipe was not pressure tested. The bypass was then mounted on temporary scaffolding poles, which allowed the pipe to twist under pressure. The bypass pipe diameter was also smaller than the reactor flanges, and so in order to align the flanges, short sections of steel bellow were attached to each of the bypass ends.

During the late afternoon on June 1, 1974, a 20-inch bypass system ruptured, which may have been caused by a fire on a nearby 8-inch pipe. This resulted in the escape of a large quantity of cyclohexane, which formed a flammable mixture and subsequently found a source of ignition. At about 4:53 P.M. there was a massive vapor cloud explosion that caused extensive damage and started numerous fires on the site. The control room roof collapsed, killing 18 people. The fires burned for 10 days after hampering the rescue work. On that Saturday, a total of 28 workers were killed, 36 suffered injuries, and 53 off-site injuries were reported. The entire facility was destroyed along with damage to private property.

Ralph King, a scientist, believed that the cause of the piping rupture was more than a simple mechanical failure: "His proposed failure mechanism was the problems started with a process upset caused by the addition of water to the reactors and a failure of the agitation system" (Sutton, 2011, para. 7). The investigation that followed the explosion found major flaws with the plant.

According to Sutton (2011), following this accident, many of the now well-established Process Safety Management (PSM) techniques were developed, especially the first paper on HAZOP by Lawley in 1974. Since this accident, and preceding and succeeding accidents, the U.S. Occupational Safety and Health Administration (OSHA, 2000) promulgated a new standard, "Process Safety Management of Highly Hazardous Chemicals, 1910.119—containing the requirements for the management of hazards associated with processes using highly hazardous chemicals to help assure safe and healthful workplaces." Many organizations have complied with the new regulations, striving for excellence at PSM.

PSM increased public concern over safety in industrial plants and led to the creation of the COMAH (Control of Major Accident Hazards) regulations, the development of Process Safety Management programs, and Process Hazard Analysis and Emergency Management. Today, PSM programs continue to improve the chemical and process industries. Although PSM is regulatory in the United States, it is still not mandated across many developed countries. Indeed, Canada, the next-door neighbor of the United States, continues to treat PSM as a voluntary requirement. Moreover, PSM is yet to be embraced as a regulatory requirement in many developing nations. Nevertheless, the authors are convinced that as the attributes and value contributions of a sound PSM program continue to become more visible, more nations will make PSM mandatory and regulated, thereby contributing to the value maximization premise of this hugely beneficial program.

Operationally Disciplined Organizations

In this section, we review some case studies that examine how organizations have found tremendous success from being operationally disciplined.

Case Study: DuPont

Founded in 1802, DuPont was started by Eleuthére Irénée du Pont to manufacture gunpowder in the United States from the advice of Thomas Jefferson, the third U.S. President. E.I. du Pont was experienced in the gunpowder production after apprenticing with the famous French chemist Antione Lavoiser. Du Pont, being awaré of the dangers of this industry, established the foundation of the company around safety. He designed and laid out the company's plant, along the Brandywine River, spacing buildings away from each other to limit the effects of any explosion.

Safety was the culture of the company. As cited by Klein and Vaughen (2008), E.I. du Pont advised, "The safety of the farmers who live in our neighborhood, has imposed upon us the absolute duty of making choice of steady, sober men and of establishing the most rigid discipline among our workmen" (p. 58). All accidents were extensively and carefully investigated; recommendations and remediations were implemented. Klein (2009) pointed out that the lessons by DuPont from the early years in the company's history are well reflected today in the many elements of DuPont's current Process Safety Management philosophy. This philosophy followed DuPont over 210 years of existence. It is also the driving force behind the company's progression to Process Safety Management and operational discipline and then excellence.

Following an accident in 1965, which resulted in 12 fatalities, 61 injuries, and over $50 million in property loss, DuPont's management began a comprehensive review of the operating and safety procedures at each of its facilities. This formally introduced Process Safety Management to DuPont. By 1979, guidelines were issued and the purpose as described by Klein (2009) was to:

- Prevent serious, process-related incidents that might affect plant personnel, off-site communities, the environment, or result in significant property loss or loss of business.
- Establish a framework to help focus management efforts on this important, serious, and complex subject.
- Comprehensively describe the principles and essential features of Process Safety Management for use by sites in managing process safety.
- Describe corporate business responsibilities and activities.

DuPont then recognized that written procedures and instructions were not always followed by employees. DuPont reflected on its main strength of

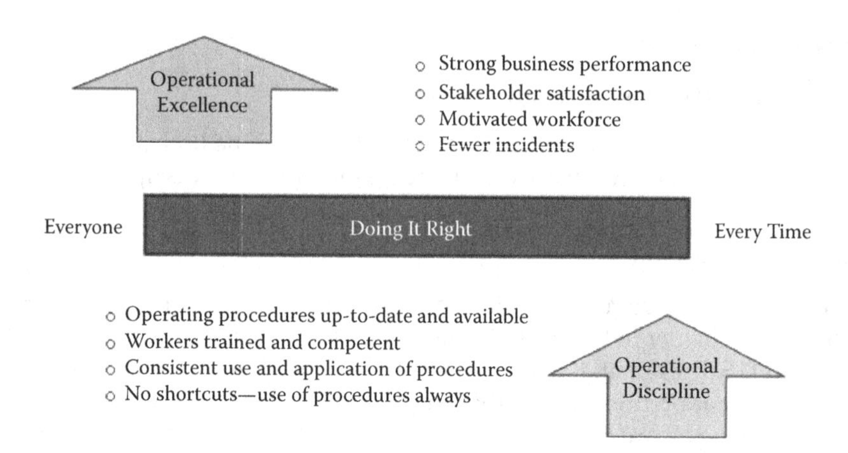

FIGURE 2.1
DuPont's operations discipline and operating excellence model.

safety culture and the benefit operational discipline (OD) in safety would add to the efficiency of its safety, health, and environmental plans. In 1989, it added OD to its PSM training plan. Initially, operational discipline was defined by DuPont as the deeply rooted dedication and commitment by every member of an organization to carry out each task the right way, and each time by doing every task the right way, every time. DuPont's operating philosophy is demonstrated in Figure 2.1.

Subsequent continuous improvements led DuPont to drive at the individual level, positioning ownership as: *I am committed to working safely as well as to simplify the existing organizational OD effort* (Klein and Vaughen, 2008). DuPont acknowledged that individual workers possess "the knowledge, commitment, and awareness to complete their task correctly and safely every time" (Klein and Vaughen, 2008, p. 58). This new approach to OD placed ownership in the hands of its most valuable assets—its workers.

To emphasize operational discipline, the company formulated a simple eight-step approach. The steps are illustrated in Figure 2.2 and each step is described next to provide the reader a better understanding of the model.

Step 1: Be Convinced Operational Discipline Is Needed

DuPont rationalized that once the business can be affected by human safety errors, then a stronger operationally disciplined environment must be created. If not, poor quality, process, and cost issues would be obstacles to achieving the business's goals. Operations discipline can be extended further in today's complex business environment. Success in today's business environment demands that all businesses maintain some form of operations discipline to remain competitive and profitable.

Indeed today, businesses are faced with more expensive and difficult to find sources of raw materials and business inputs, increasing environmental

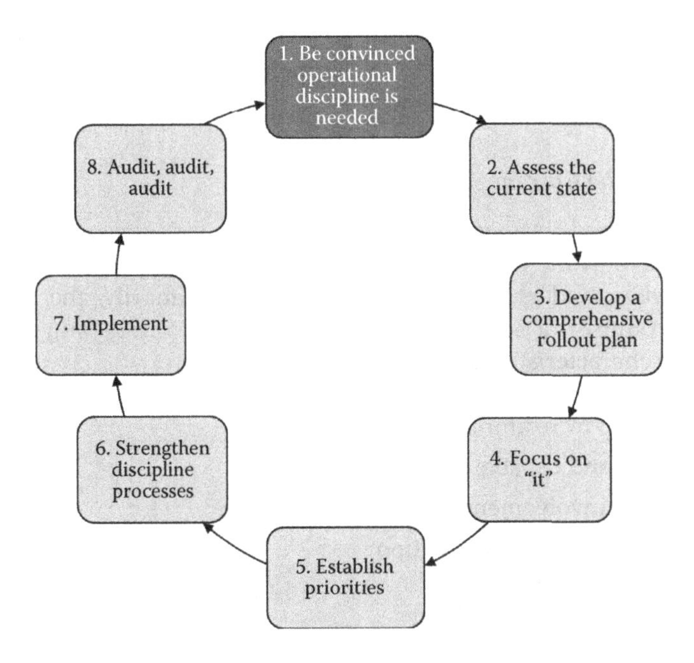

FIGURE 2.2
DuPont's eight-step approach to operational discipline. (From Rains, B. D., 2010, Operational discipline: Does your organization do the job right every time? Retrieved July 24, 2012, from http://www.humanerroranalysis.com/Operating-Discipline/Operating-Discipline.pdf. With permission.)

protection and human welfare pressures. As a consequence, organizations are hard-pressed to operate without some form of an operations discipline model. The challenge for most organizations is a clear definition of what the requirements are for operations discipline in each business area or industry.

For many process organizations, such as energy, oil, gas, and chemical, and in particular large organizations, operations discipline is established to varying extents in these organizations. Operations discipline may be found in pockets of the business, in business units, or business areas, and may not always be a key element of the organizational culture. Key drivers for determining if operations discipline is required in your business lies in the responses to the following questions:

1. Does the business process have the potential to injure or kill workers?
2. Can the business process result in damage to the assets and the environment?
3. Are we as efficient as we can be in our business?
4. Are we operating in a business that is challenged by declining or scarce and expensive inputs?

5. How will the image of the business be impacted from major events resulting in fatalities, spills, environmental contamination and damage, and human welfare concerns?

Step 2: Assess the Current State

DuPont developed various diagnostic tools to assess the operational discipline culture in the company, including a self-assessment methodology, online surveys, and characteristics assessment. Primarily, the diagnostic measures comprise a set of 11 characteristics. Klein and Vaughen (2008) listed these characteristics as follows:

- Leadership by example
- Sufficient and capable resources
- Employee involvement
- Active lines of communication
- Strong teamwork
- Common shared values
- Up-to-date documentation
- Practice consistent with procedures
- Absence of shortcuts
- Excellent housekeeping
- Pride in the organizations

Rains (2010) advised "an assessment team provides site management with feedback for each of the 11 OD characteristics. Strengths and improvement opportunities are highlighted, as are best practices at other sites" (p. 2). As the OD plan was improved and with the implementation of an individual OD, these characteristics were simplified and classified as organizational characteristics. DuPont found that many of the 11 characteristics were overlapping, which led to amalgamation and further classification into four core focus areas:

- *Leadership focus*—Passionate about safety and model the expected behavior.
- *Employee involvement*—Active participation in safety activities.
- *Practice consistent with procedures*—Completed tasks following the authorized procedures.
- *Excellent housekeeping*—Maintaining equipment and work areas in good operating condition.

Three characteristics were developed for promoting OD at the individual level:

- *Knowledge*—Understanding the proper procedures for various tasks.
- *Commitment*—Doing the tasks right every time.
- *Awareness*—Recognizing potential problems and unfamiliar situations.

Today, these seven characteristics are applied at DuPont in assessing, evaluating, and improving the overall OD plans.

Step 3: Develop a Comprehensive Rollout Plan

Developing a comprehensive rollout plan is no different than planning for any business process. As with any business process, having a detailed plan leads to reduced problems during execution of projects. "An OD improvement plan must be consistent with the company's values and management must be credible and authentic in its communication and implementation" (Rains, 2010, p. 2). It is essential that any changes are documented in the OD plan and properly stewarded.

Change management is an important component of any rollout effort. Change management as opposed to management of change (MOC) refers primarily to stakeholder impact from any proposed change. Careful planning for rollout is essential for success. Since people impact is greatest from any change, it is critically important that the rollout process addresses the change management requirements of all stakeholders (Table 2.1). Change management therefore requires the following:

- All stakeholder groups be properly identified.
- Each stakeholder group engaged to properly understand the impact of the change on the stakeholder.
- Actions and activities required to mitigate the impact of the change for each stakeholder group.
- Stewardship of actions to ensure change impact is properly managed.

TABLE 2.1

Sample Change Management Template That Addresses the Needs of All Stakeholders

Stakeholder Group	What Is Changing	Impact of Change	Mitigation Strategy	Who Is Responsible	Status	Date Completed
Stakeholder 1						
Stakeholder 2						
Stakeholder *n*						

Step 4: Focus on "It"

Step 4 follows the main theme of doing the task the right way every time. It involves a technical review, employee involvement, procedure writers, and a document management system. DuPont viewed this as essential to build the culture of operational discipline. To achieve the required focus attention, DuPont's strategy involved providing documented operating procedures that were easily accessible to all workers. This strategy ensured that employees did not follow the wrong procedure and obtain a bad result, or completely ignore management directives, or even revert to doing the task the best way they knew how, thereby inviting negative consequences and suboptimal performance.

Step 5: Establish Priorities

All procedures and actions must be listed according to priority. Following the precedence set by E.I. du Pont, a list of critical safety rules must be highly visible for all employees and contractors to be fully aware of and comply to. In 1811, E.I. du Pont issued official safety rules and a statement (as cited in Klein, 2009, p. 115): "As the greatest order is indispensable in the manufacturing as well as for the regularity and security of works, than the safety of the workmen themselves, the following rules shall be strictly observed by every one of the men employed in the factory." DuPont assumed a zero tolerance stance on all willful violations.

Step 6: Strengthen Discipline Processes

Operational discipline needs to be consistent throughout the organization; consequently for employees who do not follow the procedures there must be consequences to these actions.

Typical steps in a progressive discipline system are:

1. Verbal warning
2. Written warning
3. Probation
4. Suspension with or without pay
5. Termination

Today's literature promotes the application of positive reinforcements as a means for strengthening behaviors. Rewarding the right behaviors is generally more effective than punishing the wrong behavior.

Step 7: Implement

For step 7, the plan and expectations must be clearly communicated to all stakeholders. Due dates should be established and met. Feedback should be solicited and a work environment that promotes employees' trust and engagement to generate value-added feedback of the change impact is absolutely essential. According to Rains (2010, p. 3), among the "most important considerations are management consistency, exhaustive two-way communication and prompt response to questions or concerns."

Step 8: Audit, Audit, Audit

Step 8 revisits the initial assessment and audits the execution of the OD plan. From this audit, further improvement opportunities will be decided upon. The same assessment tools are used from step 2. The purpose of this audit is to determine if true progress was generated by comparing performance after the change was executed to the situation prior to the change.

Case Study: Chevron

Chevron Corporation is among the forerunners in the integrated energy companies and is based in California. Chevron's business activities span the entire globe. Its success is primarily attributed to the operational discipline and loyalty of its 62,000 employees and 200,000 contractors. A truly integrated organization, Chevron wears the hat of producer, transporter, refiner, marketer, and distributor of crude oil and natural gas. Much like its peers, Chevron is now also exploring the development of renewable energy sources and biofuels. Chevron is also involved in generating power, creating geothermal forms of energy, and producing and marketing petrochemical items.

Analysis of these high-risk production systems and dangerous commodities goes hand in hand with ensuring that proper safety measures are adhered to across all levels of the organization. PSM is deeply rooted in Chevron's business. Operational excellence delivers protection of the environment and communities while still being a reliable and efficient company. Incidents are viewed as preventable, and the governing policies, monitored processes, implemented tools, and needed attitudes and behaviors are instilled to ensure this.

Operational excellence has many components, which include the adherence to five major objectives:

- Seeking and achieving an injury-/incident-free workplace.
- Providing a workplace that is healthy with the elimination or reduction of health risks.
- Maintaining high awareness and prevention of environmental and process risks.

- Ensuring high levels of efficiency, integrity, and reliability are maintained in the company.
- Being mindful of the usage of resources and assets to limit wastage.

When considering the first objective of running an incident-/injury-free workplace, as health and safety practitioners we acknowledge that this does not happen overnight. A cultural shift in the organization is required to achieve this goal. Leveraging training, safety practices, and procedures, and constant upgrade, Chevron moved closer to achieving this goal. Creating a culture that embraces these ideals through self-awareness of every employee and sustaining these ideals are essential to sustain this achievement.

Providing a workplace that is healthy with the elimination or reduction of health risks emphasizes the importance of assessing, removing, and reducing health risks. Considering the nature of the business and the risk of not just incident and or injury, risks may arise from chronic or acute health hazards. Personnel exposure to resources, fuels, and gases can be very dangerous. Risk assessment methods help limit the possible sources of danger to employees. The ability of the company to recognize hazards and immediately take action to rectify them is one of their strong points. There is no space for imperfections or shortcomings when it comes to their adopted safety regime.

Maintaining high awareness and prevention of environmental and process risks requires adherence to safety standards and work standards to maintain safe processes and operations on all levels. Dealing with highly volatile and dangerous materials requires utmost levels of care and standardization. Training of all parties, which includes suppliers, contractors, and employees, has to be consistent and reliable to ensure the highest levels of awareness so that environmental spills and loss of containment are avoided.

Ensuring high levels of efficiency, integrity, and reliability are maintained in the company highlights the importance of integrity, efficiency, and reliability in the company's business operations. Stakeholders must identify Chevron as a preferred business partner and consumers must feel comfortable in business relationships with Chevron in order to continue choosing them above other producers and suppliers for products and services. Maintaining these essential business relationships require Chevron to maintain the reliability and efficiencies similar to that of the business stakeholders.

Being mindful of the usage of resources and assets to limit wastage requires doing it right the first *time*. Operational excellence behaviors and business processes lead to conscious and smart choices by Chevron in the usage of resources and asset management. A conscious effort toward value maximization drives the organization toward business decisions that are best for not just the company's benefit but the best for the world.

Case Study: Exxon Mobil

In 1989, the Exxon Mobil's company approach to safety changed dramatically due to the fallout from the 1989 *Valdez* oil spill. The accident triggered what is now known as a comprehensive framework called its Operations Integrity Management System (OIMS). This management system extended beyond PSM to guide discipline and excellence in every operating and business decision the establishment now makes.

A key outcome of the *Valdez* oil spill was the realization that safety extended further than being a business priority. Rather, safety must be integrated into business decisions in every decision-making policy in the organization at all levels to ensure value maximization. To ensure that this vision was to be a reality in the day-to-day running of the company, Exxon proceeded on a top-to-bottom review of all operations with the intent of reorganizing the company to embed safety in its culture for the protection of its people, equipment, and the environment. The goal was to make Exxon Mobil accident and incident free. To achieve this goal, Exxon Mobil developed and implemented its OIMS to provide strategic direction at all levels of the organization to meet its safety commitment and value maximization objectives. Exxon's OIMS framework consists of 11 elements with an underlying principle and a set of expectations. These are summarized next.

Element 1: Management Leadership, Commitment, and Accountability

It is significant that the first element of the OIMS is management leadership and accountability. Exxon Mobil managers not only comply with the elements laid out in OIMS, but they are also expected to lead the process by demonstrating a visible commitment to safety and element 1 requirements as an example to every single employee and contractor working for the company.

Managers establish policy, provide perspective, set expectations, and provide the required resources for successful operations. At Exxon Mobil, safety leadership and commitment starts at the top and is reflected in the behaviors of every employee of the organization. Managers are evaluated by their overall performance in complying with OIMS requirements. By doing this, Exxon Mobil reinforces the importance of putting safety and operations integrity first.

Element 2: Risk Assessment and Management

An important step in managing risk is proactive identification and management of the particular risk. Through a structured and consistent process, Exxon Mobil identifies and mitigates potential risks and hazards to all personnel, facilities, the public, and the environment. Steps were designed and implemented to mitigate those risks and evaluate the successes and weaknesses. Decisions are clearly documented. This protocol extends to the entire company.

Element 3: Facilities Design and Construction

At the very foundation of safe operations is sound design and construction. All Exxon Mobil facilities, new or modified, must meet or exceed applicable regulatory requirements. OIMS requires the following specifications before any operation is executed:

- Construction is in accordance with specifications.
- Operations integrity measures are in place.
- Emergency, operations, and maintenance procedures are in place and adequate.
- Risk management recommendations have been addressed and required actions taken.
- Training of personnel has been accomplished.
- Regulatory and permit requirements are met.

Element 4: Information and Documentation

OIMS requires that Exxon Mobil facilities maintain accurate and current information on their configuration and capabilities, as well as all material safety data sheets (MSDS). This is to facilitate the upstream and downstream chemical hazards that may be posed. Information and documentation includes access to updated procedures and policies.

Element 5: Personnel and Training

Exxon Mobil employees understand that Operations Integrity Management is part of their job. Employees are trained to identify hazards and assess risks with their jobs. This ensures that workers voice any safety issues they observe, be it operational or physical hazards, their own behaviors, or the behavior of their coworkers.

Element 6: Operations and Maintenance

OIMS calls for key operating procedures to ensure Exxon Mobil's facilities are operated within established parameters. In particular, these procedures address critical equipment, which must be tested by qualified personnel to ensure that the equipment poses no danger to the operator or any personnel. An effective and disciplined preventative maintenance program is maintained to ensure high reliability and operations excellence.

Element 7: Management of Change

When incidents occur, they are often preceded by a personnel, process, physical, or asset change that is often not properly managed. Maintaining safety and operations integrity over the life of a project requires having a system in place that allows for successfully managing change in operations, procedures, standards, facilities, or organizations. In its OIMS framework, all Exxon changes, temporary or permanent, must be evaluated to ensure operations integrity risks arising from these changes are addressed.

Element 8: Third-Party Services

Third-party contractors represent a significant part of Exxon Mobil's global workforce, so it is important that they understand and meet the requirements established by the OIMS. For this reason, Exxon Mobil invests significant time and effort to ensure contractors understand expectations and are appropriately trained to carry out their responsibilities. The selection of third-party providers is heavily influenced by their ability to work in a safe and environmentally sound manner. Once contractors have been selected, their work is continuously monitored to ensure it conforms to Exxon Mobil's standards and all regulatory requirements.

Element 9: Incident Investigation and Analysis

One of the most important aspects of OIMS is the reporting and analysis of safety and operations integrity incidents and near misses. Any incident of significance, even one that may not result in harm or injury, is investigated and documented, and steps are taken to help ensure such an incident does not reoccur. Findings are shared with workers who could benefit from them. Lessons learned from past incidents outside the company can also play a major role in continuing to strengthen and foster good safety management practices and performance.

Element 10: Community Awareness and Emergency Preparedness

The OIMS requires a specific set of actions to ensure that, in case of emergency, all appropriate steps are taken to protect the communities in which they operate and the surrounding environment, as well as company personnel and assets. It requires operations to have detailed plans in place to respond rapidly to incidents, and to test and continuously improve through simulations and drills.

Element 11: Operations Integrity Assessment and Improvements Evaluation

Exxon Mobil operations are regularly assessed to see how well they satisfy OIMS expectations. The more complex or high risk the operation, the more frequently it is assessed. Every assessable unit, such as a refinery, chemical plant, or production operation, performs an OIMS self-assessment annually. In addition, every 3 to 5 years, units undergo an external OIMS assessment. In an average year Exxon Mobil performs 40 to 50 external OIMS assessments at its operations around the world.

Case Study: BP

BP is among the largest private-sector energy corporations in the world. It is a vertically integrated cartel that operates oil and natural gas exploration, marketing, and distribution all over the world (Kamiar, 2010). BP has been involved in multiple major environmental and safety incidents. The Texas City, Texas, refinery is BP's largest and most complex refinery with a rated capacity of 460,000 barrels per day and an ability to produce up to 11 million gallons of gasoline per day. The isomerization (ISOM) unit is used to convert raffinate, a low octane blending feed, into higher octane components for unleaded regular gasoline. The unit has four sections including a splitter, which takes raffinate and fractionates it into light and heavy components. The splitter consists of a surge drum, fired heater reboiler, and a fractionating column 164-feet tall.

On March 23, 2005, an explosion and fire at the Texas City refinery claimed the lives of 15 workers and injured many more. The explosion and fire occurred after personnel responsible for the startup overfilled the raffinate splitter tower and overhead contents resulting in overpressuring of its relief valves. Liquid was pumped into the tower for almost 3 hours without any liquid being removed or any action taken to achieve the lower liquid level required for startup procedure. At least 20 times more liquid was pumped into the tower without activation of the automatic liquid level control as mandated in the startup procedure (Chappell, 2005).

A decision late in the startup to begin liquid removal from the tower exacerbated the incident, as rapid heat exchange between the overheated liquid being removed from the bottom of the tower and the liquid feed continuing to enter the tower caused significant vapor generation within the tower and out of the unit, overpressuring the relief valves and ultimately overwhelming the adjacent blowdown unit (Chappell, 2005). The investigation team estimated that about 1100 barrels of liquid was discharged to the blowdown unit, which had a capacity of 390 barrels. Most of the liquid was released into the petroleum sewer system. An estimated 50 barrels overflowed the tower and led to the formation of a hydrocarbon vapor cloud at ground level. The source of ignition was not known.

On January 16, 2007, the BP U.S. Refineries Independent Safety Panel released the report of its investigation of the safety culture at BP's five United States refineries and was based on the Texas City explosion. Investigations into the incident revealed the following critical factors that led to the explosion and greatly increased its consequences:

- Process safety, operations performance, and systematic risk reduction priorities had not been set and reinforced by management. In fact, BP had not adequately established process safety as a core value across the U.S. refineries. BP emphasized personal safety, which was done well, but not process safety.
- The working environment was resistant to change due to lack of trust, motivation, and a sense of purpose. Unclear expectations between supervisory personnel and management led to consistent breaking of rules.
- A poor level of hazard awareness and understanding of process safety resulting in workers accepting high levels of risk.
- Poor vertical communication and performance management process. BP had not provided effective leadership on or established appropriate expectations regarding process safety performance. Executive management either did not receive or did not effectively respond to information that may have indicated that there were process safety deficiencies in some facilities.
- Unit supervisors were absent from the scene during critical parts of the startup, and unit operators failed to take effective action to control the process upset or to sound evacuation alarms after the pressure relief valves opened.

What was interesting was that the panel concluded that the report findings should not be limited to BP, but be used as a guide for every process-based industry in the United States due to the lack of a process safety culture in many established companies.

As a result of the unpleasant attention brought to the operations side of the business due to the Texas City refinery explosion, BP approached the Massachusetts Institute of Technology (MIT) to assist it in improving its operations. MIT had assisted BP in 2003 to strengthen and standardize the way BP managed its billion-dollar-plus projects via Projects Academy, an award-winning professional education program. BP acknowledged "with the benefit of hindsight, we didn't always spend enough time focusing on operations as our core activity and discipline and recognizing our operations community to be as important as it is in such a technical and complex business (MIT News, 2008).

Through the MIT Professional Education Program and the Sloan Executive Education office, the new Operations Academy (OA) was designed to enhance the safety and operations capability of BP's operations leaders and to develop the executive leader's abilities to nurture a culture of continuous improvement at BP. The main objectives of the OA were to:

- Create a long-term way to continuously improve their operations through people and processes, and understand how the principle of "defect elimination" is applicable to all aspects of BP's operations,

helping first to meet the essentials of risk mitigation, legal compliance, and BP basic requirements and then toward efficiency and operations excellence.
- Listen to and empower the frontline as agents of change.
- Establish more systematic and rigorous work methods through the establishment of the Operating Management System (OMS as the common language and framework across BP's operations).
- Enhance organizational and leadership skills to ensure sustainability of efforts.
- Build a strong community of operations leaders across the company who own and continue to develop BP's approach to operations.
- Deepen the technical capability of leaders at all levels with particular emphasis on Process Safety Management.

Senior managers were required to improve their technical and management skills at MIT and assist in establishing the company's OMS as the BP way of operating across its global operations. The OMS was expected to provide a framework of processes, standards, and practices to help deliver consistent performance, and progress to excellence in operations and safety. OA graduates were prepared to lead a more systematic and rigorous approach to the management of operations utilizing continuous improvement as the sustainable foundation of technical integrity, reliability, and safety (Ernst & Young, 2010).

In 2007, BP's chief executive officer and 25 senior staff launched the 3-day Operations Academy Executive Program, designed to educate top management on how to lead BP to operations excellence in an enterprise that employs over 96,000 people in over 100 countries. Following the Texas City refinery explosion in 2005, a six-point plan was drawn up to address immediate priorities for improving Process Safety Management and operational risk management across BP's operations worldwide.

The six-point plan referenced commitments on the following:

1. Removing occupied portable buildings in high-risk zones, and blowdown stacks in heavier-than-air, light hydrocarbon service.
2. Completing all necessary major accident risk assessments.
3. Implementing group standards for integrity management and control-of-work on a locally risk-assessed and prioritized basis.
4. Improving the way in which they seek to ensure their operations maintain compliance with health and safety laws and regulations.
5. Closing outstanding actions from past audits.
6. Ensuring the competency of teams in matters of safety and operations (Ernst & Young, 2010).

In 2010, BP experienced one of the world's greatest disasters with its Gulf of Mexico spill. In its *Deepwater Horizon* project something went drastically wrong. Eleven workers were killed, more than 4 million barrels of oil were estimated to have been released into the Gulf of Mexico, and there was a huge economic impact and environmental fallout to the surrounding states. This incident demonstrated that OMS was yet to permeate the entire BP organization.

Case Study: Bahrain Petroleum Company (BAPCO)

The Bahrain Petroleum Company (BAPCO) is one of the largest and the oldest in the Middle East, dating to 1936. BAPCO refines over 270,000 barrels of crude every day. BAPCO senior management is of the opinion that superlative business performance starts with extraordinary safety, health, and environment (SH&E) performance. Creating a safe working environment for employees, contractors, and customers is a vital part of the company's management strategy, which is geared toward one of BAPCO's ultimate goals of operational excellence.

Although not required by the law, the company fully implemented PSM by 1995. Using this approach, the process design, process technology, process changes, operational and maintenance activities and procedures, nonroutine activities and procedures, emergency preparedness plans and procedures, training programs, and other elements that affect the processes within the company were considered. Every department was encouraged to take part in the design and implementation of the PSM system. BAPCO took almost 4 years to fully implement 12 PSM elements.

The PSM elements used were:

1. Process Safety Information (PSI)
2. Process Hazard Analysis (PHA)
3. Operating Procedures
4. Mechanical Integrity (MI)
5. Safe Work Practice
6. Management of Change (MOC)
7. Training
8. Incident Investigation
9. Contractors
10. Emergency Planning and Response
11. Pre-Startup Safety Review (PSSR)
12. Compliance Audit

All new employees were required to attend a 1-day training program. In addition, new employees underwent a structured training program, which was termed *Individual Training Program* (IDP). This program was a customized training scheme based on the roles and responsibilities of individuals to eliminate general training where only some would benefit. All training sessions including IDP's were evaluated and analyzed to identify areas for improvement. This was vital for effective PSM as needs changed consistently.

In 1983, BAPCO undertook a program of Hazard and Operability (HAZOP) studies, which was aimed at improving the safety and reliability of its operations. HAZOP is currently continuing under the Process Hazard Analysis PSM element. The purpose of these studies was to assess whether the SH&E measures were acceptable in new or existing projects or plants. HAZOP was conducted across the organization for existing and new projects to ensure a complete scope of the study.

Case Study: Outsourced Process Safety Management

"Consultants are good, because they force us to do what we know we should do," according to the health, safety, and environment (HSE) director at the Hungarian energy MOL Group (DuPont, 2012, p. 1). Sometimes it is necessary to outsource your PSM development to give a company that already has done a lot for safety to shore up blind spots with a fresh perspective. This is exactly what the MOL Group did with safety consultants from DuPont.

Safety: A Prerequisite for Developing Multinationals

MOL turned private in 1995 and has developed into Central Europe's most influential player in the energy market and now has holdings in Hungary, Slovakia, Croatia, Bosnia-Herzegovina, Serbia, Slovenia, Austria, Italy, Romania, the Czech Republic, Russia, Oman, Yemen, Syria, and Pakistan. During a rapid expansion drive to become a leader in the multinational scene, MOL Group's senior management wanted its safety performance to be on par with the top percentile of its peers leading to a demonstration of the company's operating excellence.

In 2003, MOL had recorded 55 lost time injuries (LTIs) and a lost time injury frequency (LTIF) rate of 2.6, an indicator measuring LTI cases against one million hours worked. By comparison, the International Association of Oil and Gas Producers in its 2003 safety performance report recorded an average rate less than half that of MOL 1.16 LTIF among its 36 member companies. Shocking results for a multinational supplying most of continental Europe; clearly something had to be done to improve safety.

Safety at the Very Bottom

The implementation of a Safety Management System decided on by senior management was not seen as something that the wider workforce might buy into. This was because the workforce could not see the effect the Safety

Management System would have on them, since these meetings, discussions, and decisions took place solely with management. There was no buy-in from the workforce. Regular monthly safety meetings were then introduced that were attended not only by the HSE experts but also by the managing director and the local operating team. Information from the meetings was transmitted back to shop floor employees, so they were also informed of what was happening.

This led to employees being more disciplined. It was key that they understood that safety steps were there for them, that they were taking care of themselves and others. They knew that warning each other about missing personal protective equipment, for example, is not a way of whistleblowing but of looking out for each other. They have also taken part in the development of safety management steps by making suggestions, some of which are very simple but very effective, like changing safety glasses and helmets so that they fit together better. Even executives could conduct behavioral audits and when they came to visit a plant, operatives can see that they now wear safety helmets, safety glasses, and safety shoes; in other words, the same equipment the operatives themselves have to wear. That sent an important and positive message.

Stepping Up Safety

A shift in cultural attitude to safety was introduced by the consultants. They practiced what they taught others, an important point of contractors and the main organization to share the same values. MOL decided to approach the safety issues in two phases: laying the foundations for an overall shift in mindset and attitude to safety, then building on the continuous cultural change with a program focusing on a properly structured and adapted PSM system once the former had been implemented.

Rewarding Excellence

An example of how the shift in attitude to safety brought about a positive change is the performance of MOL Refining, the crude oil refining unit of MOL. In 2003, MOL introduced an annual HSE Award at which one division from the group received an award for its safety efforts and performance. In March 2008, this division was MOL Refining, honored for its impressive improvements. MOL Refining took a leading role in setting up the Safety Management System introduced by DuPont and today carries out an impressive number of behavioral audits, a staggering 800-plus in 2007.

Behavioral audits focused on a dialogue with employees about safety. These dialogues were conducted with employees working safely to acknowledge their positive behavior and with employees working unsafely to convince them of the unnecessary risks they are taking. The next step was to jointly develop a safer approach toward the work conducted. Extensive studies revealed that unsafe behavior was the main source of injuries rather than engineering deficiencies. As a result of the consequent execution of the HSE action plan, the number of LTIs at MOL Group Refining dropped from 9 in 2005 to 3 in 2007, and the LTIF rate decreased from 1.53 in 2005 to 0.67 in 2007.

Statistics Indicate a Shift in Attitude

DuPont completed its safety management consultancy engagement in mid-2007. By the end of 2007, the overall LTIF rate for MOL Group had already dropped from the 2003 figure of 2.6 and 55 LTIs to 1.52, which equated to 37 LTIs.

Why Do More?

Safety statistics at MOL Group were improving at a progressive rate, indicating a positive change in attitude throughout the organization. Many other organizations would have stopped there. However, MOL Group had always intended to support behavior safety management with a process safety program. Positive safety statistics can easily lead to complacency and a misinterpretation of safety performance, because serious process incidents happen relatively infrequently. But when they do, the outcome can be fatal and can affect not only the site but also the company's business, its customers, the environment, and the local community, that is, the company's reputation.

For this reason, process safety errors were more critical than conventional hazards, which are addressed by behavioral safety systems. Inadequate preventive maintenance, poor technical design, insufficient emergency planning, and incomplete hazard analyses are just a few of the reasons for major incidents. Process safety management addresses these and other likely causes.

Tailored Support

Key performance indicators (KPIs) to evaluate PSM at group level were implemented to push the PSM drive forward and to make proper benchmarking of all the work that had gone in already. The objective for DuPont had always been to make the MOL Group self-sufficient and give MOL Group employees the skills to carry the PSM program forward themselves. This part of the program was called Train the Trainers. During a transitional period, DuPont consultants accompanied new MOL Group process safety trainers as they rolled out the program to different sites.

By mid-2008, the MOL Group had become self-sufficient and established a large, dedicated process safety Network with more than 120 expert members. To ensure that everyone in the group knows what is expected of them, all PSM requirements have now been set out in the new MOL Group PSM Global Operative Regulation. Process Safety Management has been made mandatory for all hazardous operations, and contractors are also given a set of standard requirements they have to abide by if they want to work for MOL Group.

The most important difference noted was that people were taking action to make operations safer in divisions such as retail, which was not part of the high-risk operations. The PSM project has been judged to be so successful within the MOL Group and the company has such confidence in its efficacy that it is being extended to other companies in the group in Pakistan and Russia.

Conclusion

From the case studies highlighted and the reported incidents described in this chapter, these were catastrophic failures of complex technical systems that resulted in fatalities, serious injuries, and major economic loss. The main conclusion of all of these incidents was related to the safety culture of the organization operating the system as an important contributing cause of the incident. For many organizational leaders, the organizational safety culture is the outcome of the individual worker, team, and organizational and group values, attitudes, competencies, and patterns of behavior that determine the commitment to and consistent application of the organization's health and safety programs.

Process safety focuses on the prevention of releases of hazardous material or energy from the manufacturing process. Process safety programs emphasize the design and engineering of equipment and facilities, hazard identification and assessment, inspection, testing, maintenance, process control, management of change, procedures and training, and learning from incidents and near misses.

BP mistakenly believed that a good occupational safety performance, as measured by the commonly used safety measures such as occupational illness and injury (OII) rate, indicated that PSM systems were performing well.

However, occupational, or "personal" safety, generally considers hazards that are directly related to individual workers. Programs to manage these hazards (slips, falls, electrocution, cuts, muscle strains, etc.) are very different from those that address process safety hazards. The focus is often on worker recognition of hazards, use of personal protective equipment, provision of guards on hazardous equipment, and following safe work practices. The worker potentially exposed to the hazard has significant control over his ability to recognize and protect himself from the hazard.

Studies have shown that there is no statistical correlation between personal safety performance and process safety performance, as measured in the United States by the number of incidents reported through the U.S. Environmental Protection Agency (EPA) Risk Management Program. As stated by the BP Panel, "the presence of an effective personal safety management system does not ensure the presence of an effective process safety management system" (p. 21). A good corporate process safety culture is demonstrated by the actual performance of the PSM systems in the operating facilities; not by the filing of paper upon paper of audits and so on. Systems and procedures are important and necessary, but they do not ensure effective PSM. The systems must be real and functioning, not just paper systems.

Actions recommended by process safety reviews must be implemented; incident investigations must be used to improve the process rather than to assign blame; mechanical integrity inspections must be completed on time and corrective actions actually taken; training at all levels must be appropriate and up to date; and operating procedures must be correct, up to date, and actually used. Most important, management at all levels from the board of directors and CEO to the frontline supervisor must demonstrate leadership for process safety at all times.

References

American Institute of Chemical Engineers. (2005). *Building Process Safety Culture: Tools to Enhance Process Safety Performance*. New York: Center for Chemical Process Safety of the American Institute of Chemical Engineers.

Chappell, R. (2005). BP issues final report on fatal explosion: Announces $1 billion investment at Texas City. Retrieved from http://www.bp.com/genericarticle.do?categoryId=2012968&contentId=7012963.

Chemical Safety Board (CSB). (2005). Investigation of the March 23rd, 2005 Explosion and Fire at the Texas City BP Refinery. Presentation by the Chemical Safety and Hazard Investigation Board.

Collins, J. (2001). *Good to Great: Why Some Companies Make the Leap and Others Don't*. New York: HarperCollins.

DuPont. (2012). Developing process safety at MOL. The second building block of an effective safety culture. Retrieved December 30, 2012, from http://www2.dupont.com/Sustainable_Solutions/ru_RU/assets/downloads/MOL_Group_Case_Study.pdf.

Ernst & Young. (2010). BP sustainability review 2009. Retrieved from http://www.bp.com/liveassets/bp_internet/globalbp/STAGING/global_assets/e_s_assets/e_s_assets_2009/downloads_pdfs/Safety.pdf.

Kamiar, M. (2010). A short history of BP. Retrieved from http://www.counterpunch.org/2010/06/16/a-short-history-of-bp/.

Klein, J. A. (2009). Two centuries of process safety at DuPont. *Process Safety Process,* 28(2): 114–122. Retrieved December 30, 2012, from: http://onlinelibrary.wiley.com/doi/10.1002/prs.10309/pdf.

Klein, J. A., and Vaughen, B. (2008). A revised program for operational discipline. *Process Safety Progress,* 27(1): 58–65. Retrieved from http://www.humanerroranalysis.com/Operating-Discipline/Operating-Discipline.pdf.

Lutchman, C., Maharaj, R., and Ghanem, W. (2012). *Safety Management: A Comprehensive Approach to Developing a Sustainable System*. Boca Raton, FL: CRC Press/Taylor & Francis.

Macza, M. (2008). A Canadian perspective of the history of process safety management legislation. Programmable Electronic System in Safety-Related Applications. Symposium conducted at the meeting of the *8th Internationale Symposium*, Cologne, Germany.

Massachusetts Institute of Technology (MIT). (2008). BP-MIT program focuses on operations safety. Retrieved from http://web.mit.edu/newsoffice/2008/bp-mit-0410.html.

Occupational Safety and Health Administration, U.S. Department of Labor. (2000). Process safety management. Retrieved from http://www.osha.gov/Publications/osha3132.pdf.

Rains, B. D. (2010). Operational discipline: Does your organization do the job right every time? Retrieved July 24, 2012, from http://www.humanerroranalysis.com/Operating-Discipline/Operating-Discipline.pdf.

Sutton, I. (2011). The Flixborough disaster. Retrieved July 24, 2012, from http://sutton-books.wordpress.com/article/the-flixborough-disaster-2vu500dgllb4m-21/.

Walter, R. (2002). *Discovering Operational Discipline: Principles, Attitudes, and Values That Enhance Quality, Safety, Environmental Responsibility, and Profitability*. Amherst, MA: HRD Press.

3

Leadership Behaviors for Network Performance, and Operational Discipline and Excellence

The Role of Leaders

Leadership, when defined, refers to setting a compelling sense of direction to followers. Where operations discipline and excellence is concerned, the key tenets of leadership continue to be the same. However, leadership plays a key role in ensuring that leaders provide the following:

- Create a shared vision across the organization that influences the behaviors of followers in the direction of the vision.
- Courage to make necessary changes and establish the desired culture of the organization.
- Demonstrate behaviors that support the values of the organization.
- Act in a timely manner when opportunities are identified.
- Hold self and workers accountable for performance against goals and objectives.
- Ensure consistent use of written standards and supporting procedures, and takes action when deviations occur.
- Provide prioritization and sufficient resources.
- Establish and steward performance management.
- Is visible and available to support and drive operations discipline and excellence.
- Ensures adequate resources for oversight of work and performance management.

In this section, the authors discuss these leadership behaviors and their contributions to operations discipline and excellence.

Create a Shared Vision across the Organization That Influences the Behaviors of Followers in the Direction of the Vision

Perhaps the most fundamental role of leadership—creating a shared vision across the organization—helps in defining those followers who accept and share the vision from those who do not. The vision statements answer the question for all stakeholders: *Where do we want to be in the future?* Those who share the vision will remain with the organization, whereas those who do not will eventually seek out new opportunities that more closely align with their aspirations outside the organization.

Leaders create and share the vision of the organization to establish the new norm of the organization. Once the vision is shared and accepted by the followers, leaders are required to work to support the vision and to establish strategic goals and missions that are aligned with the vision of the organization. When a new operationally disciplined and excellent vision is created, the goal of the leader is to establish the new aspirations of the organization, and the values and beliefs necessary for achieving the vision of the organization. Indeed, Networks are created to deliver the organization's desired state consistent with the vision and that may include operations management, Process Safety Management (PSM), HSSE (health, safety, security, and environment), or Management System-focused Networks.

Attributes of a Shared Vision

Among the key attributes of a shared vision are the following:

- The shared vision uplifts the aspirations of all stakeholders in the organization.
- A shared vision harnesses and channels people's energy and excitement toward a common goal.
- A shared vision provides the necessary direction, motivation, and momentum for all stakeholders in the organization.

Once properly shared by senior leaders (CEO and executive leaders) of the organization, workers and stakeholders quickly fall in line and set about the tasks at hand to achieve the vision.

Creating a Shared Vision

Once the vision of the organization is established, the difficult part is to share the vision among all stakeholders of the organization. Sharing of the vision internally is a critical step in the process of creating a shared vision since these are the people who will ultimately enable the organization to achieve its vision. Creating a shared vision involves the following steps and activities:

- Engaging and involving key people in the organization—Namely recognized and trusted members and leaders within the organization.
- Start with leadership at the highest level—All senior leaders of the organization must share the vision. Failure to acquire buy-in and support for the vision by all senior leaders can potentially result in infighting and internal corporate politics that can be detrimental to the organization.
- Communicate, communicate, communicate, then communicate again—Creating a shared vision involves an intense amount of communication among all stakeholder groups to create a united front for the organization. Communication at various levels of the organization varies in content and granularity based on the intended audience. Engaging professional communications experts is critical in establishing the right key messages and the mode of communication to ensure the best hit for each target stakeholder group.
- Be realistic—Ensure alignment with resource capabilities (no pie in the sky). When visions are created within organizations, they must be simple, easily remembered, and aligned with the core business and resource capabilities of the organization.

Table 3.1 differentiates the attributes of strong versus weak vision statements.

It is also important that stakeholders recognize and understand the relationships among the day-to-day operations of the business (work plans), organizational mission, and the organizational vision. Figure 3.1 provides an overview of the relationships among these three commonly misunderstood components of the business activities.

Courage to Make Difficult Changes and Establish the Desired Culture of the Organization

Particularly for a mature organization, values are generally well established. As organizations seek to move to transform themselves to operationally disciplined and excellent, such moves represent a shift in the values and accepted norms of the organization. Often such changes can be quite

TABLE 3.1

Attributes of Strong and Weak Vision Statements

Strong Vision Statements	Weak Vision Statements
• Short	• Long
• Focused	• Unfocused
• Clear	• Unclear and ambiguous
• Memorable and easy to recall	• Easy to forget
• Easy to understand	• Complex and difficult to understand

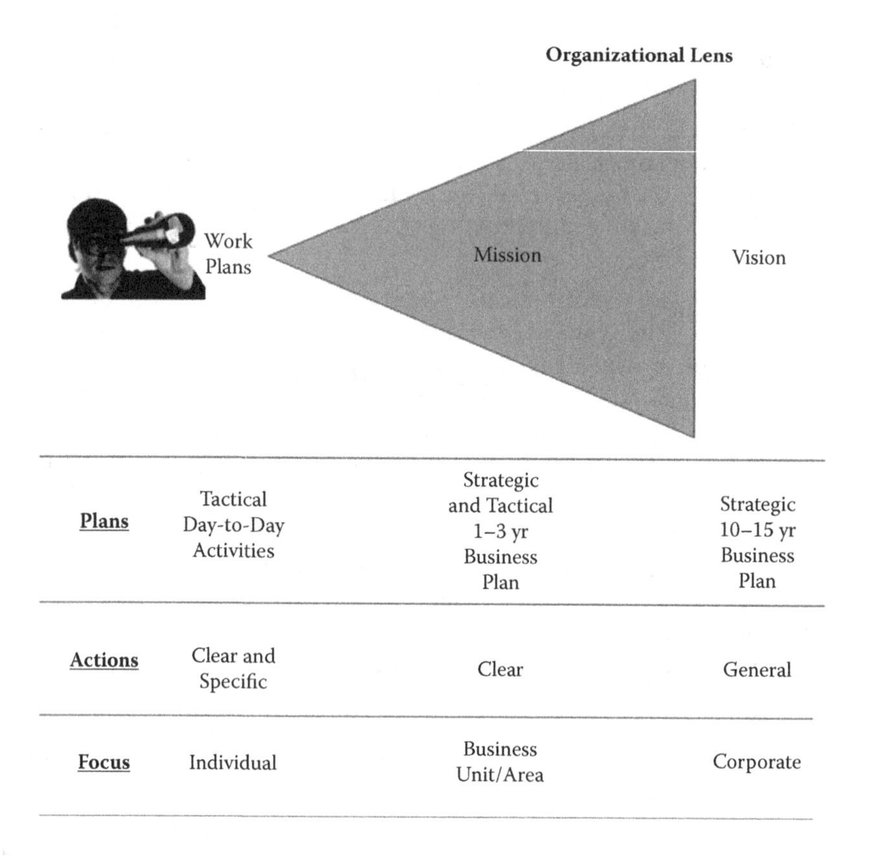

Plans	Tactical Day-to-Day Activities	Strategic and Tactical 1–3 yr Business Plan	Strategic 10–15 yr Business Plan
Actions	Clear and Specific	Clear	General
Focus	Individual	Business Unit/Area	Corporate

FIGURE 3.1
Differentiating among work plans, mission, and vision of the organization. (From https://www.safetyerudite.com/. © 2012 by Safety Erudite Inc. With permission.)

impactful to the business and require significant leadership courage to make these necessary changes.

Changes in the values and belief systems of an organization impact each worker at an individual and often personal level. As a consequence, such changes are among the most difficult to execute across the business and take a significant amount of leadership courage to execute them. Nevertheless, the value created from such courageous leadership actions, once demonstrated to all stakeholders, have the potential to generate momentum and support for the change.

Courageous leadership is essential to make any significant changes in the business. It takes extreme courage to be different and to assume leading roles in today's highly complex and competitive business environment. This is particularly so when faced with new and unconventional processes where outcomes have not yet been subjected to the challenges of industry peers and stakeholder scrutiny with differing goals and aspirations (e.g., unions whose goals may be to maximize employment and organizational focus on mechanization to improve efficiencies).

An example of courageous leadership as applied to Management Systems is in the sharing of knowledge generated from incidents within organizations. Many Management Systems today seek to promote learning from health and safety incidents from internal and external sources. Business leaders, however, have historically avoided doing so as they seek to avoid the following:

- The impression of airing dirty laundry in public.
- Concerns around liability exposures from doing so.
- A desire to avoid being pioneers and leaders in this area (there is no clear leader in the industry, we do not wish to be the first to do so).
- Going against the legal opinion; legal advice that is contrary to doing so.
- The fear of being wrong.

It is interesting that almost every senior business leader is aware that 90% to 95% of incidents are predictable and avoidable, and 80% to 85% are repeated.

Surely, if our goal is to avoid repeating costly incidents and drive discipline and excellence in our businesses, we should seek to learn from the experiences of our peers and other industries. Sadly, many of our leaders have failed to demonstrate the courage required to share knowledge in health and safety. Consequently, we continue to repeat many costly incidents within our business that could have been avoided from Shared Learning.

Demonstrate Behaviors That Support the Values of the Organization

Leaders must visibly demonstrate behaviors that support the values of the organization. Where operations discipline and excellence is concerned, they must demonstrate behaviors that inspire the hearts and minds of all stakeholders. Within their arsenal of leadership behaviors should be the following:

1. Ethical and morally sound behaviors
2. Genuine care and empathy for people
3. Reaching out and engaging workers and people
4. Fair and consistent treatment to all workers
5. Standing up for the right things; taking actions to correct nonaligned follower and peer behaviors
6. Coaching, mentoring, and leading
7. Doing the right thing
8. Being accountable

Demonstrated leadership behaviors send a clear message to all stakeholders that these are the expected behaviors to work in this organization. Such behaviors should be aimed at inspiring the hearts and minds of workers.

Act in a Timely Manner When Opportunities Are Identified

Leaders who procrastinate and fail to respond to business opportunities in a timely manner can often demotivate workers. This is particularly so when the perceived opportunity is in the area of environmental protection and improving the welfare and health and safety of the workplace. Timely decision making helps businesses take advantage of available business opportunities. The key to success, however, is to follow the internal management system of the organization in a disciplined and organized manner. When decisions are made consistent with the internal Management System processes, decisions are better analyzed and follow a consistent transparent process to the benefit of the organization.

Various personality types and tests have been historically used for identifying future leaders. Myers-Briggs and True Colors personality tests are commonly used to identify leaders who are quick decision makers. However, care must be applied when using these methods to select leaders since other desirable leadership attributes may be overlooked in favor of decisive decision making. When decisions are made on a timely basis, followers are inspired and are motivated to give more. Decisions, however, must reflect the values of the organization and should align with the strategic plan, mission, and the vision of the organization. Timely decision making is advantageous to the business when health and safety is concerned since it sends the message that we are socially responsible and we care about our workers and the environment.

Hold Self and Workers Accountable for Performance against Goals and Objectives

Once corporate goals are established in the annual and strategic business plan, leadership must be accountable for the performance of all followers and take corrective actions as required. Where performance exceeds goals and targets, behaviors should be rewarded. On the other hand, where performance is less than expected, required corrective actions by leaders must be undertaken.

Corrective actions may include the following:

- Training and competency assurance to upgrade the skills and behaviors of workers.
- Reallocation of work if assigned work was too much.
- Reassessment of personnel resources to ensure proper work–life balance in execution of work.

- Reexamination of goals and objectives to ensure they are SMART (specific, measurable, achievable, realistic, and time-bound).

Accountability for performance deliverables should be an ongoing task. Effective performance evaluation is done several times during the year so that people are aware of their ongoing performance.

In mature organizations, personnel are held accountable for deliverables on an ongoing basis. Workers are subject to performance on the following frequencies:

- Quarterly,
- Semiannually, and
- On a major activity basis (after each assigned project).

Furthermore, some organizations adopt a self-performance evaluation with peer and leader oversight and review. Accountability and performance management on an ongoing basis helps to drive discipline and excellence in the business.

Lutchman, Maharaj, and Ghanem (2012) outlined goals translation and alignment across all layers of the organization and across all stakeholder groups for EH&S (environmental health and safety)-related performance deliverables. Figure 3.2 shows the application of the same model for all business and corporate deliverables beyond EH&S deliverables. According to Lutchman et al., workers respond better to corporate goals and objectives when they know what needs to be done and in some instances how the work is to be done.

Once the corporate goals and objectives are established, they must be communicated across the entire organization very quickly. All leaders must seek to establish a line of sight or alignment between the works they do relative to the corporate goals. In addition, leaders must ensure their followers are knowledgeable about how the work they perform contributes to the corporate goals and objectives.

Ensures Consistent Use of Written Standards and Supporting Procedures; Takes Action When Deviations Occur

Operations discipline and excellence is a natural outcome of the consistent and continuous application of written standards and supporting procedures for all critical and nonroutine work. When standards and procedures are followed, costly mistakes and errors are avoided. Lutchman et al. (2012) highlighted roles and responsibilities of the various stakeholder groups regarding use of procedures and standards (Figure 3.3).

Lutchman et al. (2012) advised, for each element of the PSM or the management system, leadership has a responsibility for establishing a supporting

Develop and communicate corporate vision	**Corporate Leadership**
Develop and communicate strategic goals and annual business plans	**Site Leadership** • Ensures annual goals are aligned with corporate mission • Develop tactical strategies for achieving annual business plan or mission • Provide oversight and stewardship (e.g., trend analysis) to identify improvement opportunities and resource prioritization
Participate in developing and communicating tactical plans for achieving short-term goals and objectives	**Middle Managers** • Develop stewardship tools for collecting data and measuring progress relative to short-term goals and objectives • Identify and close gaps and may prevent achieving short-term and business plan goals
Supervise and manage work safely to ensure short-term goals and objectives are met	**Frontline Supervisors and Workers** • Ensure work is conducted in a way so as to ensure tactical plans are executed safely and goals and objectives are met • Ensure work-plan contain tactical measures and strategies to achieve short-term goals • Promote continuous improvements in safety and worker performance
Supervise and manage work safely to ensure short-term goals and objectives are met	**Contractors** • Ensure work is conducted in a way so as to ensure short-term safety goals and objectives are met • Ensure work-plan contain tactical measures and strategies to achieve short-term goals • Promote continuous improvements in safety and worker performance

FIGURE 3.2
Alignment of goals and objectives to achieve strategic objectives. (From Lutchman, C., Maharaj, R., and Ghanem, W., 2012, *Safety Management: A Comprehensive Approach to Developing a Sustainable System*, Boca Raton, FL: CRC Press/Taylor & Francis. With permission.)

standard to which the organization must adhere. Leadership is also accountable for developing the supporting procedures for conformance to the organizational standards. Standards and procedures developed must be simple to apply and must deal with all requirements of the element such that workplace safety can be properly addressed.

A core requirement of standards and procedure management where leadership is concerned is enforcing use and continuous application to the point of automatic responses. Variances in use and application of standards and procedures may be required for the following reasons:

Develop and maintain corporate standard	**Corporate Leadership**
Develop and maintain supporting procedures	**Site Leadership** • Define and approve template for procedure • Ensures procedure alignment with Standard • Identify site roles and responsibilities in applying the procedure • Drives compliance in use of procedures • Recommends improvements to the standard as required
Enforcement in use of procedures	**Middle Managers** • Ensures all personnel are trained and competent in the use of the procedure • Ensures procedures are accessible to all users • Verifies procedures are correct and upgrades when deficient in meeting safe working conditions
Consistent use of procedures	**Frontline Supervisors and Workers** • Provide oversight in the use of established procedures • Ensure all workers are trained and competent in the use of the procedures • Promote continuous improvements in the application of the procedure
Consistent use of procedures	**Contractors** • Provide oversight in the use of established procedures • Ensure all workers are trained and competent in the use of the procedures • Promote continuous improvements in the application of the procedure

FIGURE 3.3
Roles and responsibilities of various levels of leadership. (From Lutchman, C., Maharaj, R., and Ghanem, W., 2012, *Safety Management: A Comprehensive Approach to Developing a Sustainable System*, Boca Raton, FL: CRC Press/Taylor & Francis. With permission.)

- Difficulties in use and executing the procedure or standard.
- Procedure may not be appropriate for assigned work.
- Technology has changed and the procedure is no longer valid for the work.
- Specialized work that may be contracted and lies outside the core competence of the organization.

Where variances are necessary, leadership is required to assess, analyze, and mitigate all risks so that work can be conducted safely and in an operationally disciplined manner.

Provide Prioritization and Sufficient Resources

Prioritizing and resourcing work activities are key leadership requirements. Using consistent practices for prioritizing is a key requirement for operations discipline and excellence. For strategic projects, organizations may establish threshold financial indicators to facilitate the go/no-go decision. Common financial indicators used in prioritization of projects may include the following:

- Internal rate of return (IRR)—A minimum net present value of 10% may be established by organizations for determining whether to invest in a project.
- Payback periods—Some organizations may establish minimum payback periods for selecting among investment projects.
- Net present value (NPV)—Organizations may establish minimum NPV for projects based on the capital invested.
- Combinations of the above—In most instances, no single measure is used for investment projects. Most organizations may use a combination of two or more of the aforementioned indicators.

Where opportunities are identified by Networks, a common tool used for prioritizing work is the opportunity matrix (OM), as shown in Figure 3.4. The OM compares the value created from project options with the complexity of executing the various projects or initiatives being compared to allow work to be prioritized. As shown in the OM, a list of activities is identified. With inputs from Network members, a score is allocated for the value created from each activity. Similar inputs from the Network helps in determining a score associated with the complexity in executing the particular project. This practical and simple tool applies nonexact science and is highly effective in supporting project and work activities prioritization in the business.

Ultimately, the goal is to land on realistic score combinations for value creation and complexity for each project or activity. Using these score combinations, projects and activities are plotted onto the radar and compared for prioritization. Those with the best value creation (high) and complexity (low) are prioritized higher in the project selection and resourcing list for execution.

Hale (2009) listed the following criteria for determining the complexity of execution of each project or activity:

- Time to implement solutions
- Cost of solutions
- Regulatory risk
- Man-hours required
- Assumptions about the technical difficulty
- Previous failed attempts

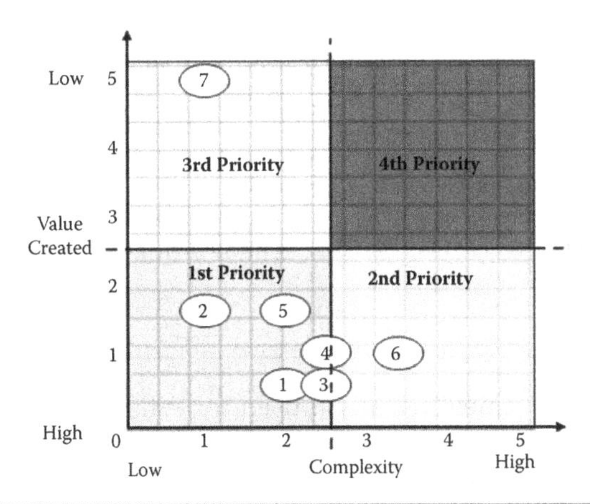

Priority	Opportunity	Complexity 1=Low 5=High	Value Created 1=High 5=Low
1	Get prequalification rolled out right	2.0	0.5
2	Align contract language with PSM/OMS requirements	1.0	1.5
3	Standardized Computer-Based Contractor Orientation (3 levels)	2.5	0.5
4	Simplify CM101 and make BU/BA specific (workshop with tools)	2.5	1.0
5	CM101 support to BU in rollout of Contractor Safety	2.0	2.0
6	Provide access to organization's work related information (procedures, policies, processes) to all contractors	3.5	1.0
7	Update the Contractor Safety Standard based on improvement opportunities identified	1.0	5.0

FIGURE 3.4
Application of the Opportunity Matrix in prioritizing work associated with Contractor Safety Management. (From https://www.safetyerudite.com/. © 2012 by Safety Erudite Inc. With permission.)

Similarly, the Network may consider both quantitative and qualitative attributes of each project to determine the value created for each opportunity. Table 3.2 shows a few of the common quantitative and qualitative variables that may be identified and addressed when evaluating and prioritizing Network projects.

Establish and Steward Performance Management

Leadership establishes the reporting requirements of key performance indicators (KPIs) to ensure continuous improvements and to assess the efficacy of processes and tools that may be in place to support the work of Networks

TABLE 3.2

Quantitative and Qualitative Measures in Determining Value Creation

Quantitative Variables	Qualitative Variables
• Project cost and dollar savings • Lost or differed earnings and production from not proceeding with project or activity • Injury consequence or environmental impact if EH&S related • Wastage and rework cost • Legal and liability cost	• Morale impact • Worker turnover impact (hiring and training time) • Brand and image impact

and continuous improvements. Criteria used for establishing and stewarding such KPIs are as follows:

- Leading and lagging indicator focus—Both leading and lagging indicators are valuable to the organization. Leadership must identify and focus on the right combinations of leading and lagging indicators for effective stewardship.
- Emphasis on leading indicators—Leaders should focus on leading indicators since they are often preventative and are early indicators of potential problems and opportunities for the business.
- Monthly reporting and stewardship—Generally a frequency that avoids overtaxing the resource capabilities of the organization is adopted. *For high impact KPIs, a higher reporting frequency may be adopted.*

Figure 3.5 provides a model for developing KPIs for the business. The model suggests that Networks created for each element or operations management requirement should focus on developing leading and lagging indicators for stewardship. Stewardship of these indicators helps to drive operations excellence in business. Leading and lagging indicators should be established for each element of the management system and packaged for stewardship under the categories of:

- People
- Processes and systems
- Facilities and technology

A more detailed process for developing KPIs for each element along with criteria for selection is provided in Figure 3.6.

Critical to the long-term success in KPI stewardship is the identification of a manageable number of high value KPIs. Focused attention on these KPIs until they are institutionalized in the business is required. As the business recognizes and accepts the value derived from these KPIs, new ones are introduced into the folds of those stewarded. Stewardship requires leaders

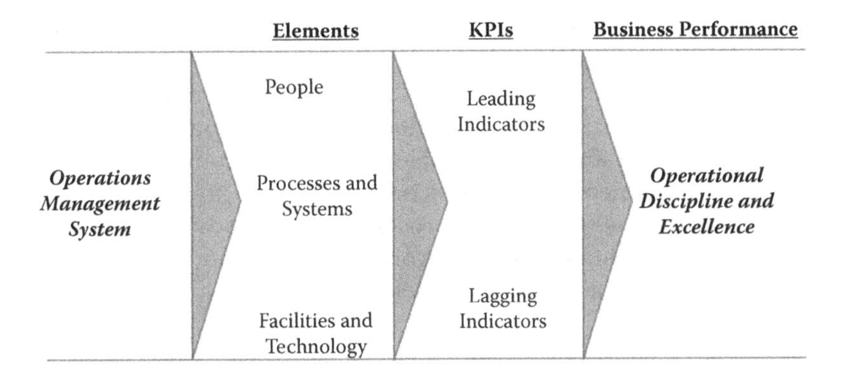

FIGURE 3.5
Model for developing and stewarding KPIs. (From https://www.safetyerudite.com/. © 2012 by Safety Erudite Inc. With permission.)

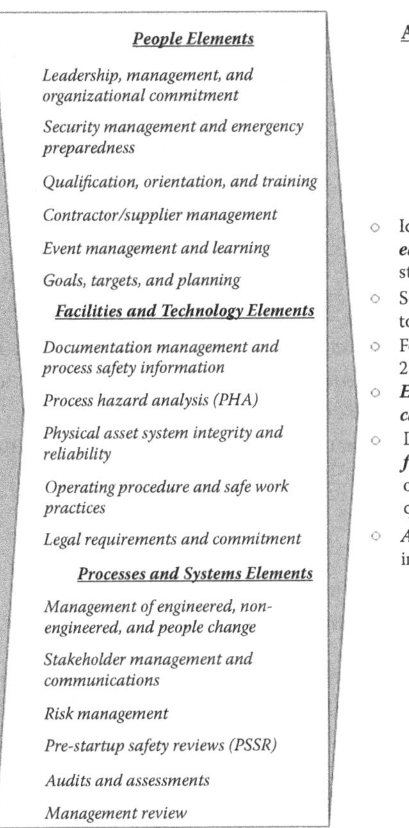

Approaches to KPIs

- Identify ***3–5 KPIs for each element*** for stewardship
- Select ***high-value KPIs*** to focus on
- Focus on a maximum of 2–3 KPIs to steward
- ***Ensure SMART capabilities*** for each KPI
- Determine the ***right frequency*** for reporting on KPIs — monthly vs. quarterly
- ***Avoid taxing the business*** in reporting requirements

FIGURE 3.6
Criteria for developing KPIs for each OMS element. (From https://www.safetyerudite.com/. © 2012 by Safety Erudite Inc. With permission.)

to establish acceptable limits for each KPI. Corrective actions are required once the KPI falls outside the acceptable norm.

Over time and with the right levels of engagement, involvement, and communication, more KPIs are introduced into the stewardship process. Eventually, KPIs that are more valuable than others from a value creation perspective are retained, whereas less rewarding KPIs are refined and even discontinued from the stewardship process. Prominent and visible displays of trended KPIs helps motivate the workforce based on performance and should be encouraged across all stakeholder groups.

Is Visible and Available to Support and Drive Operation Discipline and Excellence

Leadership visibility has a profound impact on motivating and inspiring the hearts and minds of all workers. More important, leaders that visibly demonstrate commitment to the corporate values and strategic goals and objectives of the organization are more effective in driving Network effectiveness and operations discipline and excellence. Leadership visibility requires discipline to do so. Leaders need to extricate themselves from the office and be visible in the field to interact with the workforce and provide necessary and required guidance. A visible leader is available to reinforce desired behaviors and work practices. Visible leaders are also able to provide guidance to correct and upgrade deficient skills and capabilities.

Periodic unannounced and planned field visits and review of work-related documentation helps to discourage shortcuts by workers, and drive disciplined use of tools and work management processes. More important, such behaviors usually demonstrate that the leader cares about worker welfare and is prepared to intervene when unsafe work or work that deviates from procedure is being undertaken. The authors are firm advocates that leadership visibility helps improve safety performance and Network effectiveness.

Ensures Adequate Resources for Oversight of Work and Performance Management

A requirement of leadership is to ensure adequate oversight and supervision is provided to all workers at all levels of the organization. Adequate worker supervision is required based on the cultural environment and the maturity of the workforce.

Oversight and supervision help to achieve the following:

1. Conformance to standards, procedures, and established work practices.
2. Use of personal protective equipment (PPE), particularly if the PPE is uncomfortable or cumbersome to use.

3. Creation of a motivated and harmonious work environment where workers are inspired to achieve *stretch* goals.
4. Shortcuts and unwanted risk taking is avoided resulting in fewer incidents.

More important, workers believe leaders are more approachable and their welfare is important to leaders when oversight is provided.

Leadership Styles and Behaviors: Impact on Safety

The authors believe that excellent process safety performance requires leaders to defend against normalization of deviation. As a workforce, we must all comply with internal standards, openly communicate our safety concerns, promote the safety imperatives, and maintain a healthy sense of vulnerability. The same philosophy can be extended to achieving superior Network performance and operations discipline and excellence in our business. We should continually:

- Defend against normalizing deviations.
- Comply with regulations and standards, and follow standard operating procedures.
- Openly communicate Management System concerns.
- Promote the management system imperatives.
- Avoid complacency and overconfidence.

Lutchman et al. (2012) highlighted the strengths and weaknesses of several different styles of leadership. They also suggested that *true leadership starts with a good understanding of the impact of your leadership style and behaviors on followers.* They also highlighted the value of transformational leadership behaviors as well as the Situ-Transformational Leadership Model for today's business environment. These approaches to leadership were selected for Network performance, and operational discipline and excellence based on the following workforce attributes:

- A more educated workforce.
- An evolving workforce comprised of both Generation X and Generation Y workers, and a growing presence of Generation Y.
- Higher expectations from leaders.
- Workers who want to be engaged.

- Increasing presence of women in the workplace.
- A workforce dominated by multiple cultures.
- An aging workforce in North America.
- A continually changing technological environment.

Transformational Leadership Behaviors

Transformational leadership behaviors are dominant in many Western workplaces. Transformational leaders seek to invoke the inner good of all workers by inspiring their hearts and minds to strive to higher levels of performance and to work safely. A key point to note: Leaders are not born; they evolve from the environment to which they are subjected and develop behaviors and styles consistent with their learning. Table 3.3 provides the strength and weaknesses of transformational leaders. These are generally learned behaviors that leverage conscious effort and practice. As with any leadership behaviors, continuous practice and effort is required to become the expert.

Transformational leaders focus on learning and applying behaviors that are essential for improving workplace performance. They demonstrate behaviors that suggest genuine care for workers. Transformational leaders generate performance across cultural, gender, generational, and geographical divides.

TABLE 3.3

Strengths and Weaknesses of Transformational Leaders

Strengths	Weaknesses
• Will create an organizational environment that encourages creativity, innovation, proactivity, responsibility, and excellence • Has moral authority derived from trustworthiness, competence, sense of fairness, sincerity of purpose, and personality • Will create a shared vision; promote involvement, consultation, and participation • Leads through periods of challenges, ambiguity, and intense competition, or high growth periods • Promotes intellectual stimulation • Usually considers individual capabilities of employees • Is willing to take risks and generate and manage change • Lead across cultures and international borders • Builds strong teams while focusing on macromanagement • Is charismatic and motivates workers to strong performance	• Leaves a void in the organization if followers are not developed to assume role

Source: Lutchman, C., Maharaj, R., and Ghanem, W., 2012, *Safety Management: A Comprehensive Approach to Developing a Sustainable System*, Boca Raton, FL: CRC Press/Taylor & Francis.

Although transformational leadership behaviors generate success in many areas, the worker maturity issue is not addressed with transformational behaviors as completely as it is in Blanchard's Situational Leadership® II.

The challenge for leaders, however, is to determine what works best, given the industry trends identified earlier and discussed. It is safe to say that no one leadership style or set of behaviors have the solution for today's complex business environment within which we operate. The authors support the earlier views of Lutchman et al. (2012) that suggests that success in leadership is derived from the following key traits, which must ultimately become learned behaviors for success in today's business environments:

1. Ethical decision making
2. Provide intellectually stimulating and challenging work to workforce
3. Teamwork
4. Training and competency of workers
5. Demonstrated empathy and care for all stakeholders
6. Involvement of all stakeholders and in particular workers
7. Fair treatment to workers
8. Flexibility and adaptability to a multiplicity of worker maturity levels and working environments

Lutchman (2010) discussed the application of situational leadership and transformational leadership behaviors for the complex project management and execution work environment under the new Situ-Transformational Leadership Model. With an ever-increasing percentage of Generation Y in the workplace, many traditional models of leadership will fail to inspire performance. More important, they have failed and will continue to fail to generate discipline and excellence in the workplace.

Situ-Transformational Leadership Behaviors: Driving Discipline and Excellence

Operations discipline and excellence require all workers to do the right thing at all times. This includes working safely, focusing on the value maximization proposition, ethical decision making, and social and environmental considerations. Operations discipline is clearly reflected in behaviors that encourage workers (with no coercion or supervision) to automatically perform the following tasks to ensure this work can be completed safely:

1. A field-level risk assessment, whereby all workers assess the risks associated with the task and apply risk mitigating strategies to remove the risk or reduce all risks to as low as reasonably practicable,

before doing the work. All workers must do this regardless of the worker maturity or the complexity or risks associated with the job.
2. Conduct prejob meetings to ensure all workers involved in the assigned task are aware of all the associated risks and the criticality of their role in ensuring the safety of the job.
3. Be prepared to stop all work if new hazards are identified while the work is taking place.
4. Look after the health and safety of peers and coworkers during the execution of any work.

These automatic and unconscious responses are a reflection of the cultural values of the worker that is developed and instilled over time.

Can you identify an organization that can boast of such automatic and culturally driven behaviors among its workers? Some may point to Exxon and DuPont. Organizations with automatically driven behaviors for operations discipline and excellence across all aspects of the business are far and few apart. What this essentially means is that there is tremendous wastage in resources and gross inefficiencies in our processes. More important, they are primarily within the control of leadership to generate changes and improvements in these processes.

According to Lutchman (2010), *leadership is about creating an organizational environment that encourages worker creativity, innovation, pro-activity, responsibility, and excellence* (p. 147). Lutchman further advised: *Leaders must create a compelling sense of direction for followers and motivate them to performance levels that will not generally occur in the absence of the leader's influence* (p. 139). To achieve discipline and excellence in the project management environment, Lutchman (2010) discussed the application of the principles of the situational leadership and transformational leadership behaviors working together under the principles of the Situ-Transformational Leadership Model. Lutchman argues that leveraging the strengths of both these approaches to leadership reflected in the right level of worker maturity (training, competency, experience, and commitment) combined with a shared vision in safety, the right motivation, and inspirational leadership, strong performance in safety management is possible. The attributes of situ-transformational leadership behaviors generate best performance when all leaders starting at the frontline understand the principles of this model and can use these behaviors when it is appropriate to do so relevant to the stage of worker maturity.

The Frontline Leader

The first point of contact between workers and the rest of the organization is through frontline leaders and supervisors. The way frontline leaders and supervisors treat workers will influence workers' motivation, commitment, performance, productivity, loyalty, discipline, and excellence. Commitment to safe work behaviors and conformance to standards and procedures starts here and is reflected in the competence and behaviors of the frontline supervisor. If frontline supervisors fail to demonstrate conformance to standards and procedures, then workers, contractors, and other followers who depend on the guidance of the frontline supervisor will fail to demonstrate discipline and excellence in performance.

All organizations are generally characterized by a workforce of varying skills and capabilities. Therefore, leaders must be able to adapt leadership behaviors to meet the requirements of the individual worker, which makes up the workforce. Ken Blanchard's Situational Leadership II provides an effective model for leading and developing workers through the various stages of worker maturity. Blanchard (2008) pointed out that when employers invested time and leadership to improve the maturity of workers, such employers experienced strong gains and productivity improvements. Table 3.4 shows the

TABLE 3.4

Leadership Behaviors with Maturity Status of Worker

Leadership Behaviors	Directing/Supporting Relationship	Worker Maturity
Directing (Stage 1)	High directing/ low supporting	Immature—Low competence and commitment. Leadership focus on: 1. Telling the worker where, when, and how to do assigned work 2. Key requirements of structure, decision-making control, and supervision 3. Primarily one-way communication
Supporting (Stage 2)	High directing/ high supporting	Immature—Growing competence; weak commitment. Leadership focus on: 1. Building confidence and willingness to do assigned work 2. Retains decision making 3. Promotes two-way communications and discussions
Coaching (Stage 3)	High supporting/ low directing	Mature—Competent; variable commitment. Leadership focus on: 1. Building confidence and motivation; promote involvement 2. Allows day-to-day decision making 3. Active listening and two-way communications and discussions
Delegating (Stage 4)	Low supporting/ low directing	Mature—Strong competence; strong commitment. Leadership focus on: 1. Promote autonomy and decision making and empowerment 2. Collaborates on goal setting 3. Delegates responsibilities

Source: Blanchard, K., 2008, Situational leadership, *Leadership Excellence*, 25(5): 19.

various stages of a worker's development and leadership focus during each stage of the worker's development.

For effective leadership at all levels of the organization, employers must therefore properly equip frontline supervisors and leaders with knowledge on the benefits of an effective management system and the requirements for each element of the management system. In doing so, organizations must also develop the skills and capabilities of frontline supervisors to positively influence the behaviors of workers for workplace discipline and excellence. The Situ-Transformational Leadership Model provides the ideal framework for developing all leaders to higher levels of skills and capabilities for influencing the hearts and minds of all workers.

According to Blanchard (2008), Situational Leadership II requires varying levels of directing and supporting behaviors from leaders based on the maturity of the worker. Blanchard suggests that leaders apply four sets of behaviors relative to worker maturity. Shown in Figure 3.7 is the application of the principles of the Situ-Transformational Leadership Model. Figure 3.7 also shows the transformational leadership behaviors required by supervisors at each stage of the worker development and the worker transitions through successive stages of situational leadership. These leadership behaviors help in developing the worker from incompetent and noncommittal to competent and fully committed.

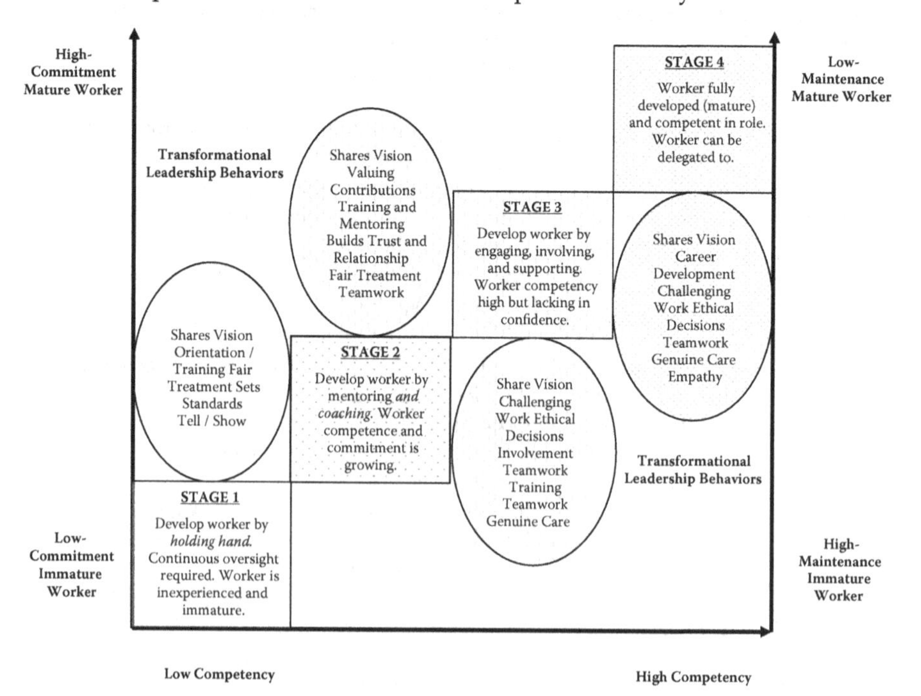

FIGURE 3.7
Situ-transformational leadership. (From https://www.safetyerudite.com/. © 2012 by Safety Erudite Inc. With permission.)

These behaviors serve to transform workers from inexperienced and indifferent of the overall organizational performance to disciplined and excellent. With the application of the Situ-Transformational Leadership Model workers are able to prioritize and exercise leadership behaviors that lead to protection of life, protection of the environment, and protection of assets, in that order, with great commitment discipline and excellence, and to the overall welfare of the organization and society as a whole.

Ultimately, the goal of any leader is to create a low maintenance worker where health and safety, and productivity are concerned. The principles of the Situ-Transformational Leadership Model help in doing so. The successful transition of workers from high maintenance to low maintenance requires that frontline leaders be adequately trained to lead the frontline workforce. Such training requires that supervisors and leaders apply the right transformational leadership behaviors at the appropriate maturity of the worker. It is noteworthy that the leadership behaviors and traits of the frontline leaders are the outcome of the training provided to them and the culture of the organization. Simply put, the organizational culture is a reflection of the way business is done, and frontline leaders are required to develop and encourage behaviors consistent with the values demonstrated by the organization.

Senior Leadership

Senior leadership must demonstrate primarily transformational leadership behaviors. According to Bass (1990), transformational leaders encourage followers to "transcend their own self interest for the good of the group, organization or society; to consider their long term needs, to develop themselves rather than their needs of the moment; and to become more aware of what is really important" (p. 53). Transformational leaders focus on developing the workforce and creating a work environment where workers feel a sense of belonging, where they are treated fairly, where they are motivated, and where they are provided intellectually stimulating and challenging work. Transformational leaders create strong teams by leveraging the abilities of experienced workers while developing less experienced workers. Lutchman (2010) advised that the "paradigm of the transformational leader is to motivate workers to do more than is expected of them by leveraging their creative excellence" (p. 147).

According to Lutchman (2010), "such leaders will create an organizational environment that encourages creativity, innovation, pro-activity, responsibility and excellence" (p. 139). They often possess moral authority derived from trustworthiness, competence, sense of fairness, sincerity of purpose, and personality. Transformational leaders are also trustworthy and ethical in decision making. Trust is an earned entity by leaders resulting from consistent demonstrated behaviors of doing the right thing at all times. In addition, transformational leaders have the unique ability to communicate and share the organizational vision. They promote involvement, consultation, and participation. Transformational leaders demonstrate high levels of emotional and cultural intelligence, and will successfully lead in volatile, fast-paced, and ambiguous work environments that are characteristic of project execution environments.

Undoubtedly, therefore, when the principles of the Situ-Transformational Leadership Model are applied by senior leaders of the organization, the impact on the workforce discipline, excellence, and productivity can be profound. Leaders who apply behaviors reflective of fair treatment of all workers, genuine empathy, teamwork, and involvement in workplace-related decisions help to create a trusting and motivating work environment. Leaders should, therefore, communicate with followers in a manner that builds trust within the workforce. Lutchman et al. (2012) advised: *Saying what you will do and doing what you said is very important in building trust* (p. 98). Workers are motivated to emulate the behaviors of leaders who make ethical and trustworthy decisions aimed at ensuring health and safety in the workplace.

Ultimately, when trust is high in an organization, workers are motivated to do the right thing at all times. They feel comfortable approaching supervisors and leaders when they are in doubt. Such behaviors help prevent incidents and drive the right levels of discipline and excellence in the business. According to The Ken Blanchard Group of Companies (2010), trust in the workplace is generated from the application of the ABCD model of leadership behaviors:

Able—Demonstrate competence

- Produce results
- Make things happen
- Know the organization; set people up for success

Believable—Act with integrity, be credible

- Be honest in dealing with people, be fair, equitable, consistent, respectful
- Values-driven behaviors "reassures employees that they can rely on their leaders" (p. 2)

Connected—Demonstrate genuine care and empathy for people

- Understand and act on worker needs, listen, share information, be a real person
- When leaders share a little bit about themselves, it makes them approachable

Dependable—Follow through on commitments

- Say what you will do and do what you say you will
- Be responsive to the needs of others
- Being organized reassures followers

According to The Ken Blanchard Group of Companies (2010), as leaders, you have a choice in becoming a trustworthy leader, as shown in Table 3.5.

The Ken Blanchard Group of Companies (2010) summarized that leaders can successfully develop organizational trust in the following ways:

1. Demonstrate trust in your people—"If you want to create a trusting work environment, you have to begin by demonstrating trust" (p. 4).
2. Share information— "Information is power. One of the best ways to build a sense of trust in people is by sharing information" (p. 4).
3. Tell it straight/never lie—"Study after study show that the number one quality that people want in a leader is integrity" (p. 4).
4. Create a win–win environment—Creating competition among workers leads to loss of trust among all.
5. Provide feedback—Hold regular progress meetings with direct reports. Check-in and provide feedback on performance, in particular, in a timely manner to avoid surprises to workers later.
6. Resolve concerns head-on—Engage workers in finding solutions.
7. Admit mistakes—"An apology can be an effective way to correct a mistake and restore the trust needed for a good relationship" (p. 5). Some cultures have difficulties admitting mistakes. However, when mistakes are admitted, it makes the leader human and promotes greater team bonding and trust.
8. Walk the talk—"Be a walking example of the vision and values of the organization" (p. 5). If the leader believes in it, then so can I and so will I.
9. Timely recognition of positive behaviors—Recognizing and rewarding positive behaviors in a timely manner. Such recognition must be specific and relevant. Choosing the right environment for doing so is also very important since some cultures may require pomp and show, whereas others may seek conservatism.

TABLE 3.5

Leadership Behaviors That Build or Erode Trust

Erode Trust	Builds Trust
• Lack of communication • Being dishonest • Breaking confidentiality • Taking credit for others' work	• Giving credit • Listening • Setting clear goals • Being honest • Following through on commitments • Caring for your people

Source: Lutchman, C., Maharaj, R., and Ghanem, W., 2012, *Safety Management: A Comprehensive Approach to Developing a Sustainable System*, Boca Raton, FL: CRC Press/Taylor & Francis.

Conclusion

Many of the traditional leadership models deliver some level of control and business performance. Indeed, the cultural environments within which business operate also help to influence the leadership culture and ultimately the operations discipline that exists with the organization. For Western cultures that have adopted transformational leadership behaviors in the workplace, the goal is to inspire the hearts and minds of workers to do the right thing at all times.

In today's multicultural societies and global workforce that is characteristic of many multinational corporations, leadership is challenged to lead and to inspire a culturally diverse workforce. Operations discipline and excellence comes from the application of consistent standards, policies, procedures, work practices, and an inspired workforce.

Developing and making available consistent standards, policies, and procedures are the easy requirements and contributions to operations discipline and excellence. The difficult component is the creation of a workforce in the hearts and minds of workers so they are motivated to do the right thing at all times. Inspiring the hearts and minds of workers is the outcome of leadership. What this means is that operations discipline and excellence requires a leadership team that understands how to do so. Enforcement as a means of generating discipline will work up to an extent and is not sustainable in the long run. Once supervision and enforcement is absent, the workers will revert to doing things the way they prefer doing it, and this may involve failing to conform to standards, policies, procedures, and work practice requirements.

The Situ-Transformational Leadership Model provides a framework for developing and inspiring the workforce based on the maturity of each worker. This model results in a trusting work environment where the workforce understands and shares the vision and values of the organization, thereby creating a transformational culture where all workers are inspired to do the right thing at all times. The outcomes of the Situ-Transformational Leadership Model, when applied correctly, are operations discipline and excellence.

References

Blanchard, K. (2008). Situational leadership. *Leadership Excellence*, 25(5): 19.

The Ken Blanchard Group of Companies, (2010). Building trust. Retrieved July 10, 2012, from http://www.kenblanchard.com/img/pub/Blanchard-Building-Trust.pdf.

Hale, R. (2009). Prioritization means agreeing what not to do, yet. Stroud Consulting Limited. Retrieved August 6, 2012, from http://www.stroudconsulting.com/fileadmin/user_upload/pdf/Prioritization_means_agreeing_what_not_to_do_yet_RBH.pdf.

Lutchman, C. (2010). *Project Execution: A Practical Approach to Industrial and Commercial Project Management*. Boca Raton, FL: CRC Press/Taylor & Francis.

Lutchman, C., Maharaj, R., and Ghanem, W. (2012). *Safety Management: A Comprehensive Approach to Developing a Sustainable System*. Boca Raton, FL: CRC Press/Taylor & Francis.

4

Shared Learning in Safety

Studies show that 90% to 95% of workplace incidents are avoidable and 80% to 85% are repeated. Learning from events is an effective means for reducing repeat incidents and cost while improving operating reliability and performance. Learning from incidents and knowledge generated from best practices and continuous improvements seldom make it to the frontline where such learning can be beneficial in preventing incidents. More important, Shared Learning is the last remaining low hanging fruit for sustained improvements in reliability, reducing incidents in the workplace, and improving organizational performance.

Many of these successful organizations have developed highly evolved internal processes for sharing knowledge, best practices, and learnings within the organization. However, such systems have evolved over time and with the application of considerable resources and commitment. In today's digital era, Shared Learning continues to limp along depending primarily upon basic computer technology for getting learning out to the frontline. Often, methodologies adopted include the following:

1. Internal folders with read/write access and controlled privileges—This methodology is highly administrative and limiting in the transfer of knowledge. Furthermore, controlled access provides a logistical nightmare that often leads to abandonment of the process or frustrated users.
2. Intranet located and accessible information—Often managed without a firm and approved process for posting content to the intranet. An evolving process as businesses seek to tighten control over quality of learnings shared on the intranet. Furthermore, since the intranet generally provides free access to all employees, the risk of inadvertent release of information external to the organization is generally high.
3. E-mails distributed to select end users—An effective method for getting relevant best practices and learnings to the right stakeholder in a timely manner. Despite timeliness, however, learnings tend to be retained in the e-mail systems of the stakeholder and will often go away with the stakeholder should he or she change jobs in the organization. Furthermore, there is no central maintained repository of these learnings in a searchable database.

In many instances, however, these processes fall short of an effective Shared Learning system, which when fully developed is intended to ensure learning is achieved.

This chapter covers the current state of sharing lessons in the workplace, the challenges faced by organizations in doing so, and industry requirements for making it happen. Process Safety Management (PSM) and other human factor considerations have improved the reliability of many operating facilities across many different industries. Similarly, significant improvements in health and safety have also been made across most industries and are reflected in declining injury frequency trends and declining numbers of incidents, injuries, and fatalities. Nevertheless, organizational learning and safety culture continue to evolve slowly.

A model for creating a continuous learning safety culture from Shared Learning is presented. The model focuses on an effective method for moving lessons (knowledge) quickly across organizations in an organized and consistent manner to foster learning to prevent repeat and similar incidents. Furthermore, the model caters to the rapid dissemination of best practices across the organization to generate business improvements and efficiencies. This chapter provides a model for Shared Learning (knowledge) and builds upon the Kaizen process for Shared Learning. The process is owner driven and facilitates rapid organized sharing of learning, both internal and external, to the organization.

Essentials for Effective Shared Learnings

To move learnings from lessons to applied learning, the following key requirements are essential:

1. Learning must be structured and presented to cater to both Generation X and Generation Y. Both groups learn differently and their learning needs must be met for success.
2. Approaches to Shared Learning must be simple and organized for sustained improvements.
3. Learning may come from several sources that include internal learning, learning from peers (within industry), and learning from other industries as well as from sources such as research and investigations done by specialized organizations like the Chemical Safety Board (CSB).

In view of these requirements therefore, the essentials for sharing learning listed next must be addressed by businesses to generate success in Shared Learning:

- Learnings should be simple and easy to understand and apply.
- Learnings should be repeatable.
- Medium and technology for sharing learnings should cater to both Generation X and Y.
- Where learning from incidents are concerned, focus should be on the following:
  - What happened?
  - Root causes of incidents
  - Key learnings from incidents
  - Recommendations to prevent incidents from being repeated
- A corporate approach to capturing and sharing learning.
- Expert Networks are required for best practices and creative solutions for generating continuous improvements (experts/subject matter experts).
- Focus on Shared Learning should be on proactive measures.
- Have a model for generating knowledge and learnings.
- Develop model for sharing knowledge and learnings. Sharing tools and processes must be user friendly, searchable, accessible to all workers, accommodating to collaboration, and secure.
- Tools for transferring knowledge and learnings to the frontline should be easy to use.
- Standardized templates and processes for sharing. Alerts, investigation summaries, and best practices help in improving the Shared Learning process.
- Establish an organizational process approval for controlling quality of learning generated.
- Action management stewardship and follow-up is required.

Current State of Generating and Sharing Knowledge

Historically, and even presently, lessons are learned after the event has occurred within the organization, essentially, a lagging indicator focus. For many organizations (even in large, deep-pocket organizations), generating and sharing knowledge from incidents continues to be weak and disorganized. Often we fail to develop internal systems and processes that can lead to the development and full sharing of knowledge generated from within the organization. A summary of the market where Shared Learning is concerned is demonstrated in Table 4.1.

TABLE 4.1
Market Summary of Current Knowledge Sharing

Sharing of Knowledge	Market Situation
Within organization, internal (business units and functional areas)	Early stages of development—Organizations are in the early stages of assessing the value of Shared Learning and are now evaluating how to share knowledge more effectively. There are no standardized processes for sharing of knowledge. Emphasis continues to be on sharing of data and information as opposed to knowledge that is beneficial to organizations.
Within industry	Disorganized and indiscriminate—Sharing within industries is disorganized and indiscriminate. Information is shared informally among peers via e-mail in a disorganized manner with no concerns of the type and quality of information being shared. There is very little regard for sensitivity of the information being shared as well as the accompanying liability in sharing.
Across industries	Not available—Very little, if any, sharing of knowledge occurs across industries.

In many instances, leaders tend to adopt protectionist approaches to data, information, and knowledge among similar business units in an organization. As a result, communication and interactions between similar business units (business areas, operating areas) of an organization may be limited and often nonexistent within the same organization. Although this concern is challenging within the organization, it is even more difficult to address between organizations in the same industry because of weak collaboration and competition regulations.

Challenges of Getting Knowledge to the Frontline

In almost every organization and industry, knowledge is continually being generated. However, this knowledge often remains in pockets of the organization, seldom reaching the entire organization to maximize value creation for the organization. There are several reasons why this happens, among which are the following:

- Leadership capabilities
- Fear of legal and market responses
- Weak understanding and communication of the benefits of Shared Learning
- Absence of the machinery within the organization for generating learnings
- Absence of an organized method for sharing and cost issues

These challenges can be addressed once organizations are aware of them and adopt a systematic approach to resolving them.

Leadership Capabilities

Historically, leaders of business units or areas that perform better from an output perspective have enjoyed the prizes of promotions and power within the organization. As a consequence, knowledge generated with one business unit or area of the business often become treasured and not shared in other parts of the business, thereby leading to suboptimal value maximization in the organization. There are many styles and models of leadership. In many organizations, there are several leadership styles and behaviors within the same work environment because of the cultural diversity of the prevailing workforce.

Addressing the leadership capabilities and challenges requires that organizations ensure leaders possess dominant transformational leadership behaviors and skills. Some of the key leadership behaviors required include individual consideration, team building, ability to create and share the vision, ability to demonstrate genuine empathy and care for workers, and the ability to lead and manage change. Such leaders must also possess excellent communications skills and technical abilities.

Fear of Legal and Market Responses

When learnings are shared within the organization, current day technologies make it the equivalent of sharing external to the organization since such learnings can be e-mailed and print copies can be scanned and electronically distributed. As a consequence, there is a general apathy to share learnings and knowledge in a manner that allows the true value of the learnings to be communicated. This is particularly so when legal considerations are included in the process as the legal department seeks to limit liabilities and potential exposures to the organization.

Similarly, the more learnings and knowledge shared, the more the market sees this as a vulnerability in the capabilities of the organization. Although sharing of knowledge brings tremendous value in preventing incidents, fear of legal and market responses places adverse pressure on leaders to share knowledge in a transparent manner so that true organizational benefits can be generated and derived.

Weak Understanding and Communication of the Benefits of Shared Learning

Toyota Corporation has generated tremendous success from *Kaizen* (Japanese term for "improvements") within the organization. Kaizen leverages Shared Learning within the organization as a means for continuous improvements. Many organizations have embraced the concept of Shared Learning within the organization for continuous improvements within specific business functions. Nevertheless, the concepts of Shared Learning have not yet been fully integrated into the PSM world for performance improvements in PSM.

More important, organizations continue to struggle with a simple means for articulating the benefits and value–effort relationships of best practices and Shared Learning for generating PSM changes. For many organizations, unless the value–effort relationship is clearly understood from making changes associated with best practices and knowledge generated within the organization, there is generally a resistance to change. Organizations struggle to communicate knowledge in a simple format that engages workers in generating and executing change. More often, where engineering learnings and knowledge is generated, the approach taken by leaders is often if it's not broken, then don't fix it.

Resolving this issue requires simple, effective means for organizations to communicate and present the value of best practices and Shared Learnings throughout the organization. Identifying the what's-in-it-for-me is essential to encourage acceptance since sustained value is generated when people are inspired to adopt best practices and learning as opposed to being mandated or forced to do so.

Absence of the Machinery within an Organization for Generating Learnings

For many organizations there is no machinery or organized process for generating learnings within the organization. Often, learning is generated at an individual level with no process for validating and approving the rollout and execution of such learning. As a consequence, many learning opportunities are lost and we frequently lose the opportunity to maximize value from sharing learning within the organization.

Many organizations fail to recognize that the application of each Process Safety Management standard within the organization can be continuously improved upon through learning. Furthermore, when organizations adopt and execute Management Systems, there are even more opportunities for continuous improvements and Shared Learning. The absences of continuous improvements machinery within organizations lead to suboptimal learning and sharing of learnings with consequential failure to maximize value.

Absence of an Organized Method and Cost Issues

For many organizations, there is no effective method for capturing and sharing learnings at the frontline. Furthermore, any consideration of internal development and stewardship of a Shared Learning process is often deemed costly and labor intensive to develop and steward. What this means is that unless there is a compelling business case for developing and executing a Shared Learning process, very little attention is placed on Shared Learning and transferring internal learning to the frontline in a sustainable and consistent manner.

In the absence of an organized effective method for transferring knowledge to the frontline, many organizations gravitate to what works best for themselves. They gravitate to a dependence on the use of e-mails and an ad hoc approach to creating and executing learning and knowledge across the organization. The outcomes of this approach tend to be an unsustainable, inconsistent application of learning across the organization that is often suboptimal in value creation and maximization.

Maximizing Value from Shared Learning

Fierce competition, dwindling resources, and social and economic pressures demand that organizations seek to improve efficiency and avoid costly mistakes. More important, negative stakeholder impact can be high when organizations fail to learn from internal incidents and that of their peers. Shared Learning in organizations remains the last low hanging fruit for improving performance. In this section, the author addresses attributes for enhanced value maximization from Shared Learning. Attributes addressed include the following:

- A corporate approach to capturing and sharing learning.
- Create expert Networks for generating continuous improvements (experts/subject matter experts).
- An approach that caters for Generations X and Y simultaneously.
- Focus on proactive measures.
- Have a model for generating knowledge and learnings.
- Develop model for sharing knowledge and learnings.
- Tools for transferring knowledge and learnings to the frontline.
- Use standardized templates and processes for sharing (alerts, investigation summaries, best practices).
- Establish an organizational process approval for controlling quality.
- Action management and follow-up.
- Sharing tools must be user friendly, searchable, accessible to all workers, accommodating to collaboration, and secure.

A Corporate Approach to Capturing and Sharing Learning

Shared Learning must be an enterprise-wide effort such that real benefits can be derived. As such, leadership engagement, commitment, and support are absolute requirements for sustained continuous improvements from Shared Learning. For best impact, Shared Learning starts at the CEO's level and must infiltrate the entire organization. Communication sessions, town hall meetings, and focus group discussions may be required to clearly articulate the vision to all stakeholders. Furthermore, demonstrated leadership commitment through resource allocation is essential to sustain the process.

Create Expert Networks for Generating Continuous Improvements

Internal and external Networks provide a very effective means for generating learning for the organization. Generally, Networks comprise of a group of experts (subject matter experts, or SMEs) that are brought together as a team with a fixed mandate of bringing about improvements in an assigned area of focus for the business. Networks may be formal or informal; however, the modes of operation continue to be the same. From a PSM perspective, several Networks are required to generate continuous improvements from learnings. A key requirement is the need for integration among Networks such that there is continuous cross-pollination across Networks to avoid duplication of efforts and consequential wastage of resources and potential conflicts. Typically, Networks are comprised of a Core Team supported by a wider group of SMEs that interfaces further still with a Community of Practice Team.

The Core Team is responsible for converting data and information into learnings and knowledge. This team is generally comprised of senior leaders of the organization who possess the ability to analyze among alternatives for creating value maximizing best practices and learning. This team is continually engaging with the organizational SMEs seeking ways and means to improve and address persistent and new problems as well as in identifying best practices that can be shared across the entire organization. The Core Team must be empowered by the organization to challenge the status quo and to generate change in the organization. Typically, the Core Team is chartered and comprised of three to five employees who are motivated to address the tasks at hand.

SMEs are a pool of resources across the enterprise that possesses expert knowledge in the area of focus of the Network. SMEs are a key resource pool for the Core Team in providing expert knowledge to the Network. They also form the conduit for activating and supporting new learning and best practices as they are rolled out across the enterprise. The SME pool may vary in size depending on the scale of operation of the enterprise or the amounts of business units or areas the organization may have. Balanced business unit representation and voice helps in improving contributions and execution of new knowledge and learnings.

The Community of Practice Team is comprised of interested stakeholders across the enterprise who are involved in the area of focus of the Network. Members of this team are the doers and are responsible for executing best practices, new knowledge, and learnings across the enterprise. The size of this team may vary based on the scale of operation and the priority of the focus area.

An Approach That Caters to Generations X and Y Simultaneously

The approach to Shared Learning adopted by organizations must be simple and should cater to both Generation X and Y (Gen X and Gen Y, respectively) at the same time. Generation X is comprised of workers who entered the workplace in the 1980s and were born during the early 1960s through mid-1970s. Generation Y workers can be categorized as "those entering the workforce generally <30 years old and born during the mid 1970s through early 2000s" (Lutchman, Maharaj, and Ghanem, 2012, p. 21). Recognizing the learning differences and expectations between both these groups, a Shared Learning approach must be technologically accessible to meet the needs of Gen Y, yet simple enough for Gen X to print and read accordingly (Lutchman et al., 2012). Also pointed out are challenges such as:

- Field experience gap between both Gen X and Gen Y. Gen X is often significantly more experienced than Gen Y and is therefore a vibrant source of knowledge. Gen Y, on the other hand, while inexperienced, seeks engagement.
- Gen Y has a shorter attention span and therefore requires concise messaging.
- Gen Y seeks change and continuous improvements on an ongoing basis and would therefore welcome Shared Learning. Gen X prefers consistency and shuns change.
- Gen Y workers are information seekers.

Ultimately, for a Shared Learning system to work, it must meet the needs of both groups of workers on an ongoing basis, since they are both dominant groups in today's work environments.

Focus on Proactive Measures

A Shared Learning system must be proactive with a focus on leading indicators and preventative measures. Leading indicators help prevent incidents before they occur leading to fewer disruptions, and greater stability and reliability of operations. Businesses can employ a pull–push strategy for transferring knowledge from engineering to operations. The pull strategy depends on workers searching for, locating, and using knowledge generated by the organization. Today's lean business environment prevents this method from being very effective since it depends on workers having the time and opportunities to seek new best practices and Shared Learning. Experience has shown that frontline workers are working with little opportunities for searching out new information.

The push strategy, on the other hand, depends on an e-mailed notification with an accessible link sent to end users when knowledge is generated. This method is highly effective since it delivers the best practices and Shared Learning directly to the desktops of end users. No searching is required and knowledge is easily accessible for use.

Have a Model for Generating Knowledge and Learnings

Figure 4.1 provides a model for generating learnings and knowledge, and for moving this knowledge out to the frontline. Knowledge generation and application depends heavily on the incident investigation and action management and follow-up sections of the model. The model must be supported with adequately trained and competent people to generate the right learning and best practices.

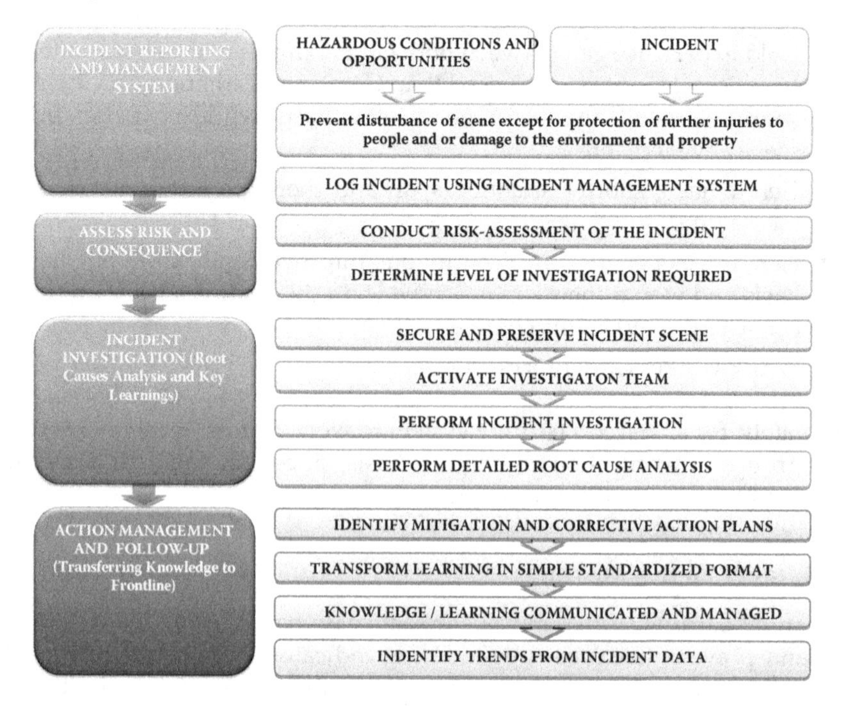

FIGURE 4.1
Model for incident management and generating knowledge and learnings. (From https://www.safetyerudite.com/. © 2012 by Safety Erudite Inc. With permission.)

Reactive Learning

Reactive learnings result after incidents have occurred and solutions are generated. Generally, reactive learnings follow an intensive investigation process whereby root causes and corrective actions are identified, as per the process defined in Figure 4.1. Reactive learnings are costly and often mandatory to the business.

Proactive Learning

Proactive learnings result when learnings generated from hazardous conditions and incidents are transferred across all levels of the business to prevent new or repeat incidents from occurring. By leveraging the Network capabilities, solutions are generated and proactively executed across the organization to improve process operations. A value–effort relationship is often used to determine whether the organization may choose to act on such learnings (Figure 4.2). Proactive learning also occurs when industry learning opportunities are identified and applied within the organization before an incident can occur.

Develop a Model for Sharing Knowledge and Learnings

Sharing knowledge and learnings requires an enterprise-wide approach for getting information to all parts of the business in a timely manner. Once learnings are generated, particularly those required for high-risk gap closure situations and opportunities, such learnings must be made available very quickly with the right level of urgency for action. A simple user-friendly approach that leverages technology is essential.

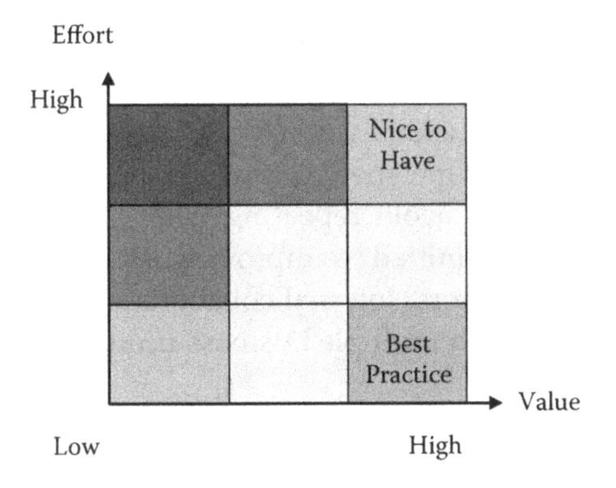

FIGURE 4.2
Value effort matrix for acting on knowledge and learnings generated. (From https://www.safetyerudite.com/. © 2012 by Safety Erudite Inc. With permission.)

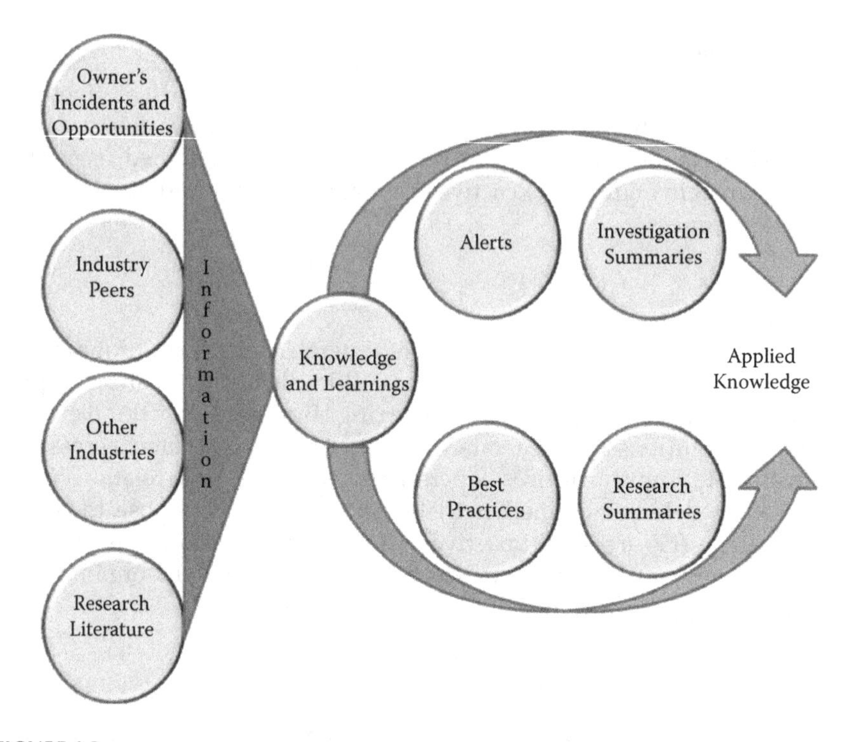

FIGURE 4.3
Model for sharing knowledge and learnings. (From https://www.safetyerudite.com/. © 2012 by Safety Erudite Inc. With permission.)

The model should cater to accessing learnings within the organization, within the industry, across industries, and from other research sources. Figure 4.3 provides a simple model for sharing based on the sharing needs identified earlier.

Owners' Incidents and Opportunities

Learnings from owners' incidents and opportunities are internal to the organization. From a PSM perspective, owners must seek to continually learn from all incidents to avoid repeating them and to enhance reliability. Owners should be committed to improving the health and safety of all workers, including contractors. Internal collaboration is essential when the organization is diverse with multiple business units or business areas, and geographically dispersed.

Industry Peers

Industry peers is a rich source of knowledge and learnings for improving PSM, to the extent that this knowledge is shared in an organized manner with the right levels of quality assurance. Where PSM and the health and safety of workers are concerned, industry peers should work collaboratively to share knowledge and learnings to protect people, the environment, and assets. An organized approach to Shared Learning is required. Current informal approaches to sharing should be replaced by organized approaches so that value creation and maximization is derived.

Other Industries

Many industries are closely interrelated and PSM knowledge and learnings from one industry can be extremely beneficial and valuable to another. For example, PSM knowledge and learnings generated from Contractor Safety in one industry is generally applicable to several industries. Similarly, knowledge and learnings generated in the power industry is also very applicable to the utilities and support divisions of the oil and gas industry. Today, no effective organized method of sharing applicable PSM knowledge and learnings exists for all industries.

Research Literature

Universities and libraries provide an excellent repository of untapped knowledge in PSM and health and safety. However, with lean organizations and tightly stretched budgets, many organizations fail to tap into this rich source of knowledge and learnings. Organizations must take advantage of cutting-edge knowledge developed by these institutions and apply it proactively across the business.

Tools for Transferring Knowledge and Learnings to the Frontline

Transferring knowledge and learnings to the frontline requires the use of simple tools and processes that are easy to follow. Knowledge and learnings can be transferred to the frontline via simple tools and methods such as alerts, investigation summaries, safety moments, and best practices.

Alerts

Alerts are ideal for use during toolbox and prejob safety meetings. Alerts provide a useful way for grounding risks and hazards in reality during prejob discussions, field-level hazard assessments (FLHAs), and work preparation. Alerts are:

- Concise, single-page summaries of learnings.
- Easily posted on visible notice boards.

- Capable of providing opportunities for learning from the experiences of others.
- Enhanced by the inclusion of a picture or image, a picture tells a thousand words.

Three types of alerts are generally recommended. These are safety alerts, technical alerts, and environmental alerts.

Investigation Summaries

Incident investigation reports can be quite detailed with information and data not generally applicable to frontline personnel. Nevertheless, summarized investigation reports provide excellent learning opportunities for all workers. Investigation summaries should be:

- Timely
- Focused (root causes, key learnings, corrective actions)
- Easy to follow to provide quick understanding, and knowledge and learnings transfers

Safety Moments

Safety moments help us keep health and safety and other learning opportunities front and center in our business decisions. Lutchman et al. (2012) highlighted that among the common themes in organizations with strong environmental health and safety (EH&S) performance was the use of safety moments for learning opportunities and knowledge transfer. Organizations with strong safety cultures and EH&S performance encourage all workers to begin all meetings with a safety moment.

Best Practices

Best practices in organizations provide perhaps the greatest opportunity for sustained improvements and performance in organizations. Although many definitions exist to define best practices, generally they refer to "any practice, know-how, or experience that has proven to be valuable or effective within one organization that may have applicability to other organizations" (O'Dell and Grayson, 1998). Best practices are exemplary or successfully demonstrated ideas or activities, viewed by some as *top-notch standards* for guiding benchmarking and making comparisons. They may be classified as *ways of doing business, processes, methods, or strategies* that yield superior business performance and results. Some international health and development organizations also believe a best practice must be *innovative, sustainable, cost-effective, ethically sound, and/or superior to all other approaches* (D'Adamo and Kols, 2005).

Best practices in organizations are extremely important sources of knowledge and learnings that can drive organizations toward operational excellence and value maximization. However, in many organizations, best practices may be siloed within a business unit or business area without being shared across the organization. Best practices can be made into standard operating procedures (SOPs) for standard application across the business. They are also repeatable and possess a value–effort relationship as shown in Figure 4.2. Generally, best practices lead to competitive advantages. However, where best practices can be advantageous to improving EH&S, organizations should adopt a proactive approach for within and across industry sharing.

In any review of best practices, it is implicit that no single practice works in all situations (Hiebeler, Kelly, and Ketteman, 1998). The term "best" is constantly evolving in a world in which good practices must change to respond to new business environmental conditions. To label any practice as best, immediately raises the possibility of dissenting voices from other companies or properties within the same organization. In addition, the term "best" may suggest that there is only one way to do things. Therefore, the terms "excellent" or "successful" could replace the term "best" to avoid the disagreements that can result when too rigid an interpretation is placed on the collection of best practices (Cornell University, 2012). Given this shifting reality, some organizations prefer to talk about *good practices* or *promising practices* rather than best practices (D'Adamo and Kols, 2005; Skryme, 2002).

Benefits of Best Practices

Identifying and sharing best practices is an important way to incorporate the *knowledge of some into the work of many.* Organizational structures tend to promote silo thinking where particular locations, divisions, or functions focus on maximizing their own accomplishments and rewards, keeping information to themselves and thereby suboptimizing the whole organization. The mechanisms are lacking for sharing of information and learning. Identifying and sharing best practices helps build relationships and common perspectives among people who do not work side by side. Best practices can also spark innovative ideas and generate suggestions for improving processes, even if a practice cannot be used in its entirety. The process of identifying them can also benefit employee morale. By highlighting or showcasing people's work, employees get organization-wide recognition for their work. All companies can learn something from the best practices of others. They can either adopt them all or at least review them to the extent to which they fit into the safety climate or culture of the organization (Chemical Industries Association, 2008).

In summary, best practices can do the following:

- Help workers learn from each other and reuse proven practices.
- Spark innovative ideas and generate suggestions for improving processes, even if a practice cannot be used in its entirety.

- Incorporate the knowledge of some into the work of many.
- Benefit employee morale by highlighting or showcasing people's work.
- Improve operations at poorly performing units so that their performance more closely approaches that at the best units.
- Help build relationships and common perspectives among people who do not work side by side.
- Save money through increased productivity and efficiency.

Successfully identifying and applying best practices can reduce business expenses, increase revenue, and improve organizational efficiency, including effective use of human resources, as long as these adopting companies have the quality infrastructure in place to make the major transformations that may be required.

Key Steps in Identifying and Sharing Internal Best Practices

To share internal best practices, you first need to recognize and define them, and identify the experts within your organization who already know how to perform them. Networks are very useful in generating best practices. With this information, the business can develop and leverage Communities of Practice that support the Network in sharing the best practice within the organization and across all levels of the organization as required. According to D'Adamo and Kols (2005), the five basic steps to identify and share best practices within an organization are:

1. *Seek out successes*—After-action reviews or debriefs. During after-action reviews, participants in an activity, event, or project conduct a structured discussion of what happened and why in order to learn from the experience. They ask the following four key questions:
   - What did we set out to do?
   - What did we actually achieve?
   - What went well?
   - What could have gone better?

   The reviews are a good way to document what people have learned before they forget what happened. After-action reviews can range from a few minutes of conversation at the end of a meeting to a formal, all-day event capping a large project. Because after-action reviews ask what went well and try to get to the root of the reason, they are a useful way to identify successes and begin defining best practices. After-action reviews also seek to learn from obstacles, mistakes, and other problems. All lessons learned, both positive and negative, are documented and shared with others.

2. *Identify and validate best practices*—Internal benchmarking. Benchmarking is a common business tool that usually compares an organization's performance with that of successful competitors. The goal is to identify, understand, and adopt superior practices and processes from outside the organization. Benchmarking can be done internally, too, by comparing the performance of different units within an organization. It is especially useful when many units perform similar activities.
3. *Document best practices*—When documenting best practices, the following requirements should be kept in mind:
   - Keep potential users in mind to ensure that what you write is user-focused.
   - Understand the needs of potential users. What problems do they want to address? How do they want to learn about best practices?
   - Best practice descriptions. Give some brief guidance, but do not write an essay.
   - Do not make rules; rather, stimulate thinking, action, and dialogue.
   - Provide enough contextual information to help users understand the conditions in which a practice has worked well and why.
4. *Create a strategic plan to share best practices*—Strategies. Once the Network has captured the best practice, create a list of contact information for experienced practitioners and point to other sources of information. The Community of Practice should actively champion best practices. The Community of Practice should work at the business unit or area level (frontline) to identify and recruit the support of people who can help create demand for a best practice. The Community of Practice should identify and focus on those people in the organization who could benefit the most from a specific practice as well as promote on-the-job learning about best practices.
5. *Adapt and apply best practices*—As we all know, one size does not fit all. Best practices therefore must be adapted to the conditions and operating environment in which they are to be used and this may vary from the point from which they are generated. Guidelines around use and application of the best practice may be required for the best results.

Obstacles to Generating and Implementing Best Practices

There are several obstacles to generating and sharing best practices in businesses. The following are a few more common examples:

- *Information hoarding*—Sometimes employees are reluctant to share their methods with others. Information can be seen as a source of power and some people hoard it. A more likely reason for not sharing is reluctance to say that something is the "best way" of doing a particular task or job.
- *The not-invented-here syndrome*—Such attitudes among workers could negatively affect the adoption of a method created by a different work group.
- *Documenting and storing descriptions of best practices*—If storage is to be in written form, a database, or other shared file system, the practice needs to be described in enough detail for all to understand. Often, written descriptions are the starting point for transfer, with employees using site visits and other forms of communication to learn. Keeping best practice information current is important. Since organizations are constantly finding ways to improve processes and products, a best practice could become obsolete.

When and How to Use Best Practices

Seeking out best practices is especially important when looking for ways to improve results of important or significant processes. In today's environment of reduced spending and rapid change, identifying ways to improve effectiveness and efficiency are critical components of managerial excellence. There are many approaches to identifying and sharing best practices, ranging from a formal organization-wide initiative with staff assigned to researching, documenting, and creating a database to more informal ways such as talking at the water cooler (sometimes the most effective approach!). A middle-ground approach involves leadership identifying the desired state, determining the parameters of a process that should be studied, and then chartering a Network team to conduct the study. A sample of people involved in the process should:

- Thoroughly review, compare, and document the current process with the desired state (gap analysis).
- Identify organizations (internal and external) that have exemplary practices or processes that produce high results; understand their processes.
- Generate possible ways to improve existing processes and close gaps identified using the knowledge derived from above.
- Select changes to be implemented and develop implementation strategy.
- Execute best practice across the organization and manage changes.
- Evaluate the results of the changes; follow the plan–do–check–act principles.

Examination of practices requires investigating new ideas, activities, or processes and determining which organizations have the most effective or profitable approach. Consequently, efforts to delineate and embrace best practices are occurring in every business sector but may still yield failure. O'Dell and Grayson (1998) pointed out that many companies have excellent practices within their own organizations but are unable to transfer and share them. The inability of many companies to share practices highlights the difficulties of using and managing knowledge. Ultimately, the challenge falls under the umbrella of change management, which addresses stakeholder interests, impact, and risk mitigation. When rolling out best practices across the organization, a stakeholder assessment map is required to ensure an effective change management process is maintained.

Internal Sharing of Best Practices

Sharing of best practices internally requires leadership attention to the following within the organization. Leaders should do the following:

- Actively promote your best practices within your own organization.
- Not expect immediate results. Activities that identify and share internal best practices are not a quick fix.
- Pay attention to motivation and organizational culture. If individuals defend doing things their own way rather than sharing and learning, best practices will be slow to spread.
- Encourage people to identify and share best practices voluntarily. Never force it.
- Start with areas that are least resistant to change and build momentum from your successes.
- Focus on sharing people's experiences as well as written documents. Practical knowledge of this kind is best transferred from person to person through direct interaction.
- Not identify best practices just for the sake of it. Focus on how they can be used to improve results.
- Demonstrate the benefits and the evidence with examples and case studies. Show how a best practice has contributed to better performance.
- Build in feedback mechanisms to create further improvements. Best practices are constantly evolving.

Use Standardized Templates and Processes for Sharing (Alerts, Investigation Summaries, Best Practices)

Standardized templates for sharing knowledge and learnings drive consistency in behaviors and encourages people to focus attention on key areas of interest. To maximize Shared Learning at the frontline, alerts and investigation summaries should, at the very minimum, provide the following:

- A brief summary of the event
- A graphic or picture
- The root causes of the event
- Key learnings from the event
- Recommendations and actions to prevent recurrence
- Demographic data—Business unit or area, contact person with e-mail address and telephone number for more information, date of event, reference documentation (standards/policies, approving authority [vice president, CEO, director])

Best practices templates should similarly be standardized with some flexibility to ensure that complete articulation of benefits are communicated. Best practices should at a minimum include the following:

- A summary of the recommended practice
- The value–effort relationship
- Subject matter experts and contact person (complete with e-mail and telephone contact details)

Establish an Organizational Process for Approval and for Controlling Quality

A process must be established for approving and for controlling the quality of knowledge and learnings shared. Approval is essential since resource commitments may be required to ensure corrective actions are initiated and stewarded. Figure 4.4 provides an overview of stakeholders and activities in the approval and quality control of learnings.

Action Management and Follow-Up

For effective transference from *lessons* to *learned,* an action management process is required. This action management process should identify the following:

- What must be done?
- Who does it?
- Who is accountable for it?
- Start and completion dates.

The adage "what gets measured gets done" is important in the action management process. Where high priority learning is concerned, senior leadership stewardship is essential to ensure lessons are transformed into knowledge and learnings at the frontline.

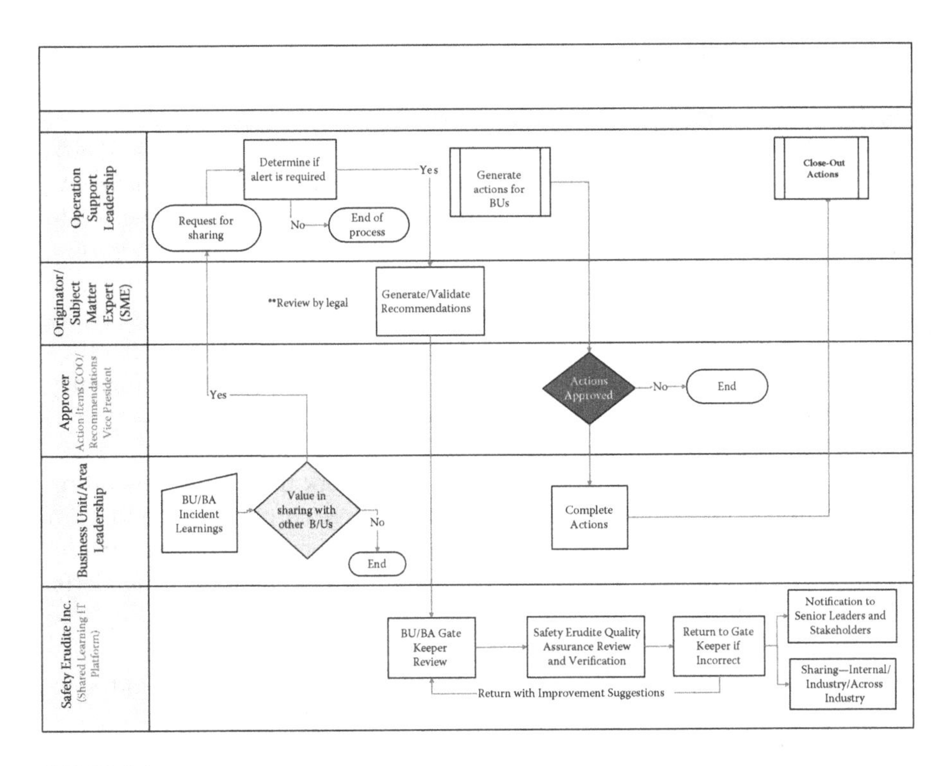

FIGURE 4.4
Organizational process for approval and quality control of knowledge and learnings generated. (From https://www.safetyerudite.com/. © 2012 by Safety Erudite Inc. With permission.)

Sharing Tools Must Be User Friendly, Searchable, Accessible to All Workers, Accommodating to Collaboration, and Secure

Very few tools exist today that provide organizations the opportunity to share learnings within the organization, within an industry, and across industries. Indeed, Shared Learning is in its embryonic stages of development. Safety Erudite Inc. (https://www.safetyerudite.com/) is a Shared Learning platform that provides organizations opportunities to share learnings and knowledge in an organized manner with strong quality control of the knowledge and learnings being shared.

Once activated within the organization, all workers may have access to the learnings. Control must be maintained over sensitive and proprietary materials. Organizations select the knowledge and learnings they may wish to share within and across industries. Typically, this is limited to safety and environmental alerts as well as investigation summaries. Sharing is typically permitted after review by the legal division of the organization that verifies no additional liabilities are brought onto the organization from sharing.

Conclusion

Studies show that 90% to 95% of workplace incidents are avoidable and 80% to 85% are repeated. What this means is that organizations are not learning from prior incidents and we must improve the way lessons are converted to learnings at the frontline. Shared Learning, or moving from *lessons* to *learned*, remains the last low hanging fruit available to organizations to improve EH&S, PSM, and Operating Performance. This opportunity is severely underutilized by organizations in the quest for improvements in reliability, reduction in the number of incidents, improvements in operating efficiency and performance, and value maximization.

With increasing stakeholder demands, leaders are challenged to find creative ways to learn. The chapter covered a Network model for generating PSM learnings internal to the organization and urges organizations to learn from its peers within its industry as well as from other industries. Learning and knowledge transfer is best achieved when simple tools such as alerts, investigation summaries, safety moments, and best practices are brought to the frontline in an organized manner and are accessible to all workers. Also identified leadership concerns and the absence of a Shared Learning process are the key constraints to improved sharing of knowledge and learnings.

A Shared Learning process and tool must be user friendly, maintain a searchable database, and be easily accessed by all workers; and be robust enough to maintain high traffic while at the same time providing the security required for the organization's knowledge. Furthermore, to maintain quality control on the learnings and knowledge shared, a gatekeeper process must be adopted.

Safety Erudite Inc. has developed an online system that promotes sharing within the organization, within the industry, and across industries, while also tapping into universities and research institutions for new knowledge and making this available to members. Complete with a push notification and action Management System as well as a collaboration room, this online resource provides opportunities to quickly transfer from lessons to learned, and to help industries significantly reduce repeat incidents by proactively learning from each other.

References

Chemical Industries Association. (2008). Process safety leadership in the chemicals industry. Best Practice Guide. Retrieved August 23, 2012, from http://www.cefic.org/Documents/IndustrySupport/CIA%20Process%20Safety%20-%20Best%20Practice%20Guide.pdf.

Cornell University. (2012). Best practices in U.S. lodging industry: Introduction. Retrieved from http://www.hotelschool.cornell.edu/research/chr/pubs/best/project/intro.html.

D'Adamo, M., and Kols, A. (2005). A tool for sharing internal best practices. United States Agency for International Development (USAID). Retrieved August 23, 2012, from http://archive.k4health.org/system/files/A%20tool%20for%20sharing%20internal%20best%20practices.pdf.

Hiebeler, R., Kelly, T. B., and Ketteman, C. (1998). *Best Practices: Building Your Business with Customer-Focused Solutions*. New York: Simon & Schuster.

Lutchman, C., Maharaj, R., and Ghanem, W. (2012). *Safety Management: A Comprehensive Approach to Developing a Sustainable System*. Boca Raton, FL: CRC Press/Taylor & Francis.

O'Dell, C., and Grayson, C. J. (1998). If only we knew what we know: Identification and transfer of internal best practice. *California Management Review,* 40(3): 154–174.

Skyrme, D. (2002). *Best Practices in Best Practices.* (K-Guide.) Highclere, England: David Skyrme Associates.

5

Creating Expert Networks for Generating Continuous Improvements

What are Networks? Historically, we have used Networks to share information and knowledge, and make changes in organizations. We have also leveraged Networks to support our personal and social interests. This chapter defines a Process Safety Management (PSM) Network as a group of experts (subject matter experts, or SMEs) that are brought together as a team with a fixed mandate of bringing about improvements in an assigned area of focus for the business.

Networks may be formal or informal; however, the modes of operation of both types of Networks continue to be the same. From a PSM perspective, several Networks are required to generate continuous improvements from learnings. A key requirement is the need for integration among Networks such that there is continuous cross-pollination across them to avoid duplication of efforts and consequential wastage of resources and potential conflicts. Typically, Networks are comprised of a Core Team, supported by a wider group of SMEs that interfaces further still with a Community of Practice Team.

In this chapter, the authors attempt to provide a conceptual overview of Networks. These concepts are explored in greater detail throughout the book. This overview explores:

- Organizational requirements for setting up Networks
- The people composition of a Network
- How Networks operate
- The various designs of Networks

Network Requirements

When organizations consider setting up Networks, there is generally a compelling business case for doing so. Triggers within the leadership for setting up Networks may include all or any combination of the following:

- Organizations have exploited common and acceptable business practices and have attained the best possible outcomes from the prevailing asset management situation. However, the potential for doing better and producing more continue to exist.
- Competitors have outperformed the organization on an ongoing basis with similar production systems and assets.
- The organization is seeking new ways of doing things to improve business performance.
- Multiple incidents in different parts of the business generate a corporate response to solving the problem.
- For the proactive and world class organization, Networks have been demonstrated as an effective means of improving business performance where they have been established in the organization.

With the recognition and acceptance of these conditions, senior leadership within the organization may seek to establish Networks to improve overall business performance. When establishing Networks, however, there are some recommendations that will enhance the overall functionality and performance of Networks. Listed next are the organizational requirements for setting up Networks:

- Clearly defined scope of work
- Business areas and stakeholder representation
- Network charter
- Support services—communications, legal, and change management
- Leadership support and sponsorship

Failure to meet these requirements may lead to a Network that performs at suboptimal levels.

Clearly Defined Scope of Work (SOW)

A clearly defined scope of work (SOW) provides the Network with a broad understanding of the magnitude of the undertaking. This SOW forms the basis of the work plan for the Network. Although senior leadership generally provides a strategic direction for the Network, a Network sponsor or sponsorship team helps the Network to clearly define the SOW that is aligned with the strategic vision and goals of the organization. For each Network that is established, a similarly aligned SOW is required. The SOW may change with changes in the business environment that may ultimately impact the strategic direction and rate of change of the organization.

Business Areas and Stakeholder Representation

Business areas and stakeholder representation in the Network is crucial for ownership and success of the Network. Involvement and engagement in developing solutions to workplace problems generally buys support for the solutions generated. Therefore, when setting up Networks, care and attention are required to ensure that key business areas and stakeholders are represented on the team. Although there is merit in limiting the size of the Core Team, representation in the SME pool of resources for all stakeholders help the effectiveness of the Network in delivering and executing meaningful solutions to the business problems so that strategic opportunities are captured and exploited.

Network Charter

A charter helps to guide the Network in terms of key deliverables and a more defined SOW. It focuses the attention on the key deliverables at hand and helps the Network in getting its work completed. Charters can vary from the very elaborate and extensive document to a simple one-page charter. For Networks, the intention of a charter is to guide the team and keep it focused on deliverables. The one-page charter is intended to minimize time spent trying to understand what was meant by each entry that is often typical of lengthy and elaborate charters.

Support Services: Communications, Legal, and Others

Although the Network may be comprised of technical and leadership expertise, there is often the need to provide specific area support to ensure success. Communications, legal, and change management support are among the essential support needs of the Network. These types of support are typical for larger global organizations where key messages must be properly crafted and the correct mode of delivery used to ensure best impact across the entire business. Similarly, legal support provides guidance to the Network in terms of regulatory compliance and legal needs for delivering change.

Expertise in change management is by far the most important type of support service that the Network may seek. The era of developing solutions, strategies, policies, and procedures in the corporate headquarters and lobbing it over the fence into the business areas for execution is over. For success in executing work in the business, stakeholders' interests must be met, and the impact of any change in the business area assessed and managed to generate the best results. Change management as a support organization will generally work with the Network to ensure a complete stakeholder impact assessment is completed and all risks mitigated such that Network-delivered changes are successful across the business.

Leadership Support and Sponsorship

In the absence of leadership support and sponsorship, Networks are destined to fail. Leadership support and sponsorship lends credibility to the work being done by the Network. This support gives the Network authority to interface with SMEs in different parts of the organization to locate and standardize best practices that can be applied across the entire organizations.

Leadership support and sponsorship also helps in removing obstacles to achieving the defined goals identified in the charter. Such support also functions as a resource enabler to the Network in securing budgets and other key requirements. Moreover, leadership support and sponsorship serves to keep the lines of communication open between the Network and senior leadership in terms of progress and general feedback.

Network Composition

The composition of Networks is critical for the overall functioning and success of any Network. When setting up Networks, leaders must pay particular attention to the following:

- Selecting a strong and credible Network leader
- Establishing a Core Team of three to eight Core Team members
- Identifying and ensuring a corporate-wide pool of SMEs
- Selecting committed, motivated team members and SMEs

When Networks are pulled together differently and without these composition requirements, such Networks run the risk of suboptimal performance levels.

Selecting a Strong and Credible Network Leader

Success of Networks depends on the leadership of the Network. Networks must be led by strong, competent, and credible leaders. Credibility in particular must be derived from past work experience and demonstrated behaviors that recognizes the Network leaders as competent and doing the right thing under all circumstances. Leaders with well developed transformational skills and behaviors that can inspire the hearts and minds of workers are best suited for Network leadership roles. Chapter 11 is dedicated to the leadership skills and behaviors essential for Network leaders since this component of Networks is very important for the overall success of the Network.

Establishing a Core Team of Three to Eight Core Team Members

When the Core Team membership of a Network is too large, there is a risk that the decision-making process can be hindered. Although it is generally recommended that membership of the Core Team be limited to three to eight people depending on the scope of work required by the Network, the optimum size of the Network is in the four- to six-member range. The key to success is ensuring business area representation with genuine interest by members in working toward a common cause for the good of the organization.

A larger Core Team helps in dividing items on the work plan so that the overall deliverable requirements are more easily met. With mature SMEs and an experienced Network Leader, managing larger teams is not a major concern. However, for the less experienced Network leader, too much time may be spent on managing the team versus focusing on and working toward the charter deliverables. Core members should be selected such that any core member can step into the Network leader's role should the leader become unable to fulfill the Network leadership role.

Identifying and Ensuring the Availability of a Corporate-Wide Pool of Subject Matter Experts

As discussed earlier, the Core Team of the Network is supported by pools of SMEs across the organization. Identified SMEs must be notified of their accessibility to the Core Team and should be released to the Core Team for short periods of time to support its work. These SMEs are generally experts in the focus areas of the Network and must be capable of sharing experience, work processes and knowledge with the Core Team in finding solutions to complex problems faced by the Network.

It is not only important to identify an available SME pool but also to ensure the availability of an extended resource pool to be drawn from as required by the Network. In many instances the business may identify and allocate SMEs without letting the SME know about the scope of work required, or the level of involvement and support to be provided to the Network. Full information and transparency is required to generate trust in the process and for securing the support of SMEs.

Selecting Committed, Motivated Team Members and Subject Matter Experts

When establishing Networks, committed, motivated team members are required. Leaders identifying these members must share the desire to achieve the strategic goal of the Network and should find competent and committed workers for the Core Team as well as to the pool of SMEs. Members must want to contribute to the cause of the Network and should not see the role as more work. Rather, Core Team members should recognize Core Team and

SME roles as prestige opportunities and a chance to provide meaningful solutions to corporate problems that hinder the organization from superior performance and best-in-class status.

How the Network Operates

The Core Team of the Network functions as would any working team. The difference, however, is that this team is likely to be faced with finding long-term solutions to strategic opportunities and solutions are not likely to be tactical in nature. As a consequence, solutions, best practices, and knowledge passed on for execution across the organization must be credible, tested, and verified either within or external to the organization.

Networks are not research teams and are not intended to research problems to find solutions. Rather, the goal of Networks is to find out what works best within or external to the organization for any defined opportunity, and to determine how best to make that process work within the organization. As a consequence, members of the Core Team should be excellent communicators and investigators so that they can work with any segment of the business to understand how the business does things in the areas of focus of the Network.

The intent of the Network is to transfer and execute knowledge to the frontline. In doing so, the following behaviors and requirements are essential:

- Focused attention on prioritized key deliverables
- Avenue created for knowledge and information transfers to and from the frontline
- Knowledge creation process; best practice identification and transfer to the frontline
- Collaboration processes

Focused Attention on Prioritized Key Deliverables

Prioritized key deliverables help the Network remain focused on the value maximization premise. As demonstrated in Chapter 8, the opportunity matrix provides an effective means for the Network to prioritize work based on complexity and value creation. Once adequately prioritized, with the assistance of the SMEs, pool resources can be allocated to completing the work plan associated with each prioritized key deliverable. As discussed earlier, mature Core Team members are required. This helps in allocating key prioritized deliverables across each Core Team member so that work can progress faster and solutions generated more quickly for the business.

Avenue Created for Knowledge and Information Transfers to and from the Frontline

Knowledge and information flow from the core Network team must be such that this can occur with very little hindrance and screening. The Network must know the real situation at the frontline that creates a real understanding of the gap between the current state of the organization and the desired state of the organization.

Networks must therefore establish the right mechanism for information and knowledge transfer across the Network (Core Team through the Community of Practice). Transparency, consideration, feedback, timeliness, and action are all important ingredients for success in this communication process. The Network must seek to identify what works best in information and knowledge transfer, and leverage that process. The support of the communication team will assist in this process.

Knowledge Creation Process: Best Practice Identification and Transfer to Frontline

When things are done well across different areas of the business, there is a sense of pride and ownership in the work practice being used. Several of these situations make the work of the Network more difficult since part of the role of the Network is to analyze these many different processes, select the best among them, and standardize for use and application across the business. For the most part, many business areas are quite responsive to improvement opportunities.

On the other hand, some business areas may be resistant to change and would provide intense challenge to the Network in executing on proposed change and capitalizing on improvement opportunities. The support provided to Networks by communications and change management personnel is invaluable in promoting adoption of new practices. A clarification of "what's in it for me" is absolutely essential for adoption and execution. Networks will have to work with each business area leveraging the Communities of Practice support to execute on any knowledge and best practices passed on to the business.

Collaboration Processes

For effective functioning of Networks, the importance of collaboration cannot be underestimated. In view of this and taking advantage of advances in communication technologies, the importance of virtual collaboration rooms cannot be underestimated. Full access to all Core Team members, read/write access to all identified SMEs, and the Communities of Practice help in creating an effective collaboration forum for generating new knowledge that is characterized by rigor.

More important, where PSM Networks are concerned—because of the overlap of many PSM elements—open virtual collaboration rooms help in facilitating communication among Networks. For success, however, where overlapping Networks exist, at the very minimum Network leads must at some agreed upon frequency review all overlapping collaboration rooms to ensure work is not being duplicated. Furthermore, such reviews may provide opportunities for bundling of solutions to be passed on to the business.

Network Design

The design of Networks is essential to the success of Networks in delivering on their priorities to the business. Typical Networks are comprised of the following:

- A Core Team
- A carefully selected group of SMEs
- An extended group or Community of Practice

Core Teams

The Core Team is responsible for converting data and information into learnings and knowledge. This team is generally comprised of senior leaders of the organization who possess the ability to analyze among alternatives for creating value maximizing best practices and learning. This team is continually engaging with the organizational SMEs seeking ways and means to improve and address persistent and new problems as well as in identifying best practices that can be shared across the entire organization.

The Core Team must be empowered by the organization to challenge the status quo and to generate change in the organization. Typically, the Core Team is chartered and comprised of three to five employees who are motivated to address the tasks at hand. Members of the Core Team should be middle- to senior-level managers who can be tasked to find solutions to business problems and opportunities.

Subject Matter Experts (SMEs)

SMEs are a pool of resources across the enterprise that possess expert knowledge in the area of focus of the Network. SMEs are a key resource pool for the Core Team in providing expert knowledge to the Network. They also form the conduit for activating and supporting new learning and best practices as they are rolled out across the enterprise. The SME pool may vary in size depending on the scale of operation of the enterprise or the amounts of

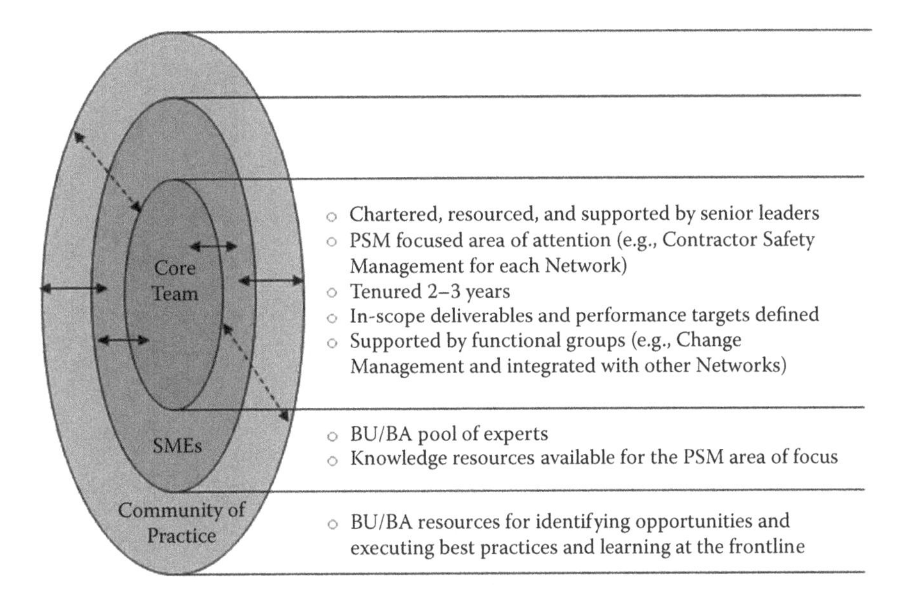

FIGURE 5.1
Sample Network structure. (From Safety Erudite Inc., 2012, Fundamentals of an operationally excellent management system, Unpublished. With approval from Safety Erudite Inc.)

business units or areas the organization may have. Balanced business unit or area representation and voice helps in improving contributions and execution of new knowledge and learnings.

Extended Group or Community of Practice

The Community of Practice Team is comprised of interested stakeholders across the enterprise involved in the area of focus of the Network. Members of this team are the doers and are responsible for executing best practices, new knowledge, and learnings across the enterprise. The size of this team may vary based on the scale of operation and the priority of the focus area. Figure 5.1 provides a sample Network structure with key stakeholder groups for generating continuous improvements and learnings.

Reference

Safety Erudite Inc. (2012). Fundamentals of an operationally excellent management system. (Unpublished.) (With approval from Safety Erudite Inc.)

6

Leveraging Networks and Communities of Practice for Long-Term Success

Companies that are striving for continuous improvement must find ways to sustain and continually improve their Management Systems. The companies that have very good Management Systems in place are working smarter, not harder. They make fewer mistakes, and when mistakes occur it is much easier to determine where and why the failures occurred. Leaders understand that a well-designed and functioning Management System will generate strong business and organizational performance. Organizations that have Management Systems in place are not fighting spot fires daily to keep their organization afloat; rather, such organizations are systematically improving the quality of their products, increasing production, improving employee morale and equipment reliability, and lowering the associated costs to produce their products.

For organizational success, leaders must see the whole picture. In most organizations leaders know that there are multiple pieces to a puzzle. However, many leaders continue to function within silos, adding inefficiency as a burden to the organization. To see the whole picture, leaders must understand the relationship of each piece of a puzzle with all the other pieces. A clear understanding of how Process Safety Management (PSM) fits into the management system helps in appreciating the big picture. More important, when leaders understand how PSM improvements impact and enhance the overall management system and organizational performance, the true progress can be achieved. PSM Networks are intended to generate continuous improvements in business. However, the way we set them up is important to the amount of benefits derived from Networks.

Once the management system is in place and the element of operation discipline is taking root in the company culture, the organization is ready to move to the next step: the creation of Networks. There are a number of activities a Network will perform, which will be discussed later in this chapter. These activities, when done well, will bring about continuous improvements to the organizational performance. In this chapter, the authors explore a simple methodology for setting up PSM Networks for generating continuous improvements in organizational performance.

Networks: A Historical Perspective

Since the earliest of times, informal Networks have existed as a means for generating improvements and social well-being. Networking may have started with the evolution of humankind. It would be hard to imagine life of early human beings on this planet without Networking (i.e., sharing knowledge, trusting each other, working as a Network to find better solutions to make life comfortable).

We all have our own informal Networks that we work with everyday, for example, if we are planning to buy a car or any other major expense item. Typically, we will connect with someone who we trust, whose advice in the past we have benefited from, and has more knowledge than us regarding that product or service. We share our opinions and knowledge with them and come up with the best plan going forward.

Today, organized and formally approved Networking is an absolute requirement in our highly competitive environment. Organized Networks must now complement the historically, informal Networks that have existed in the workplace as a means of achieving business objectives. Informal Networks, for example, are those work groups that may lie outside of the organizational structure of professional business groups that have been developed based on trusting relationships among members of the professional business groups.

There are two types of Networks: informal and formal. Formal Networks in business are in the early stages of development within organizations as businesses seek to improve processes and business practices to improve efficiency, productivity, and overall performance. Although informal and formal Networks may be characterized by different operational and organizational structures, it is important to point out that they both exhibit the following common attributes:

- Trusting relationships
- Ability to influence
- A leader with leadership skills and capabilities
- Ability to select and prioritize among alternatives (scientific approach)
- Engagement, involvement, and collaboration
- Willingness to share information
- Strong desires to improve
- Discipline
- Willingness to consider things from another Network member's viewpoint
- Ability to articulate and share the big picture
- Willingness to set and strive for aggressive goals

From a learning perspective, organizations have historically tried different approaches for addressing specific business issues in their quest for achieving best-in-class status and continuous learning. Among its people-related arsenal of tools were teamwork, committees, rapid response teams, and management Networks. Nevertheless, many of these were marginally successful and eventually faded away when leaders failed to recognize and derive immediate value.

History has also shown over time that Networks do work regardless of whether they are formal or informal. Organizations that have reached world-class status have leveraged the concepts and attributes of formal Networks. Successful Networks were characterized by leadership commitment, operational discipline, clearly defined goals, and supporting resources. To enable successful corporate Networks, they must be established with the right trained Network members, adequate Network structure, stewardship of Network activities, and sharing of learnings with users throughout the organization.

Network Creation: Essential Requirements

It is very important that an organization understands why Networks are required before embarking on the task of setting them up within the organization. Some of the requirements for setting up Networks include the following:

- A clear articulation of the value the particular Network will create for the organization.
- A business case for doing so, since Networks are expensive to set up and maintain.
- A determination of whether a permanent or temporary Network is required. For example, where the organization may have a specific issue—and once the issue is dealt with—the Network will be abolished, a temporary Network may suffice. Whereas when the organization seeks to develop best practices and continuous improvements from internal or external learnings, a permanent Network may be required.
- Composition and focus of the Network.

The focus of the authors is primarily on permanent formal Networks. The authors discuss the benefits of permanent Networks as well as the considerations required when establishing Networks.

Essential Network Criteria

What are the essential criteria for activating a Network in your organization? Since Networks are costly to set up and maintain, organizations must pay particular attention to essential criteria before Networks are created in the organization. This section defines and explores some of the essential criteria for setting up Networks. The following is a list of criteria the authors identify as essential requirements for setting up a Network:

1. What is the purpose of the Network? Can we articulate a value?
2. What is the structure of the Network?
3. What is the modus operandi of the Network?
   - How does the Network operate and communicate to achieve its goals?
4. Do we have the right leadership and subject matter expertise resident in the organization for supporting the Network?
5. Is there a clear understanding of the areas of focus and subject areas or issues to be addressed by the Network?
6. Are the goals and objectives of the Network clearly defined?
   - Is the scope of work for the Network clearly defined?
7. Are there any Network linkages with other Networks, functional groups, senior leaders, extended Networks, or community of users that are to be developed or leveraged for success?
8. What are the key success factors of the Network?
9. What is the level of autonomy allowed for the Network?
10. Is a Network charter available?

Once these requirements have been fulfilled, the organization can proceed to set up the required Network to deliver on the stated goals.

What Is the Purpose of the Network? Can We Articulate a Value?

When defining Networks within an organization, it is important that leaders are fully aware of the value creation potential to be generated from the Network. In some instances, value created or generated from Networks may be easily quantifiable and can be easily translated into a net present value for the organization, thereby fulfilling a business case requirement. For example, Networks associated with rotating equipment can easily articulate the immediate value generated from their work in the form of:

- Reliability improvements

- Reduced down/outage time
- Lost production avoided and so on

For other Networks, such as a Contractor Safety Network, it may be more difficult to quantify the immediate value created from the Network. However, on a longer-term basis, the value of such a Network will be highlighted in reduced incident numbers and declining contractor injury frequency while working at the organization's sites.

What Is the Structure of the Network?

Several formal Network structures exist in today's business environment. Inkpen and Tsang (2005) identified three common Network types:

- Intracorporate Networks—These are generally highly structured Networks among organizations.
- Strategic alliances Networks—Generally structured Networks formed between firms that enter into business relationships for achieving a common goal.
- Industrial districts Networks—Generally, unstructured Networks among a collection of firms.

The authors add to this list of Network structures to include *intercorporate Networks,* which essentially refers to a Network formed with internal stakeholders across the organization. It is particularly important to ensure all stakeholders are involved in the Network to ensure best knowledge is generated and exploited across the organization.

How Do Networks Work?

How do Networks operate? Unlike strategic alliance and industrial district Networks, intercorporate Networks are nonhierarchical and are designed to function with a fair level of autonomy within defined boundaries identified by the corporate leadership of the organization. Networks described in this book function as a group of subject matter experts who are selected by senior leaders of the organization to champion a particular cause within the organization.

Network activities and deliverables are handled along with regular duties of the worker and can often consume 20% to 25% of each Network member's time. Networks must be granted flexibility in meeting frequency and venue to ensure a balanced lifestyle is maintained by the worker. Leveraging technology in the form of live meetings, video- and teleconferencing, collaboration rooms, and instant messaging are all modes of meetings for the Network. However, there is nothing like a good old face-to-face meeting to get a lot of work done. Consequently, Networks should be mandated to hold face-to-face meetings at some interval to ensure continuity in the personal and team building component of a strong Network team.

Do We Have the Right Leadership and Subject Matter Expertise Resident in the Organization for Supporting the Network?

Network leaders must have strong subject matter expertise as well as well-developed leadership behaviors. Communication, organization, and facilitation skills are absolutes in Network leadership roles. Network leadership must be strong, focused, and transformational. Leaders must be able to create environments where members feel comfortable in contributing so that all knowledge can be tabled and reviewed for consideration and applicability. Moreover, Networks must be regarded by external members of the workforce as a prized opportunity whereby there is continuing desire by the general workforce to want to become a member of the Network. Ideally, therefore, leaders must be able to inspire the hearts and minds of its members and clearly articulate the vision of the organization as well as how the Network's work is aligned with and supports this vision.

Is There a Clear Understanding of the Areas of Focus and Subject Areas or Issues to Be Addressed by the Network?

Through collaborative efforts within the organizations, leaders must identify the areas of opportunities that the Network may be required to address. It is important that when Networks are developed, all members are fully aware of what they are getting into. A broad scope of work must be available; the time and travel commitment required and the expertise requirements must all be clearly laid out for each member.

PSM provides opportunities for Networks to be created for each PSM element within the organization. Organizational leaders sponsoring the Network must be able to identify and articulate the following to Network members:

- Gaps in compliance or opportunities to leverage
- The goals of the Network, prize for which the Network aspires
- A list of key priorities for Network focus
- Assets and leadership commitment in support of the Network
- Linkages, if any, with other Networks
- Key stakeholders who may be impacted
- A RACI (responsible, accountable, consulted, informed) chart as required

These core requirements are critical in getting the Network off to a good start to be able to deliver on priorities for the organization.

Are the Goals and Objectives of the Network Clearly Defined?

Clear goals and objectives provide direction and momentum to the Network. Goals and objectives (deliverables) can be defined in a charter for the Network. With clearly defined goals, Networks can begin work immediately toward these deliverables. In the absence of clearly defined deliverables, frustration can occur and Network teams can lose momentum in trying to determine the problem to be resolved or the opportunity to be pursued.

Once the goals and objectives are clear, a scope of work can be identified, resources can be allocated, and work can begin in order to complete the scope of work. Goals must be aligned with the corporate vision so that synergies can be identified and taken advantage of as Networks for the various PSM elements are created and begin generating value for the organization.

Are There Linkages with Other Networks, Functional Groups, Senior Leaders, Extended Networks, and Community of Users That Are to Be Developed or Leveraged for Success?

Identifying and acknowledging various other stakeholders helps to optimize performance and capture synergies. Furthermore, duplicated work is avoided, and tools and practices developed by other stakeholders can be leveraged for use by the Network, which eventually drives standardization, operational discipline, and ultimately operational excellence.

Identifying linkages and leveraging them provides opportunities for learning among Networks and for best practices transfers across each stakeholder group. More important, such linkages provide opportunities for standardization of practices, processes, and creation of a common model across the organization.

What Are the Key Success Factors of the Network?

What does success look like for the Network? How is success measured, as far as Networks are concerned? In many instances, the impact of the work done by Networks is not immediate. Moreover, it is even less obvious when qualitative measures are available for assessing performance. Organizational leaders must recognize that results may not be immediate and they must ensure that undue pressure is not placed onto Networks when performance is not easily quantified.

In view of this, as best possible, measurable success factors should be SMART (specific, measurable, achievable, realistic, and time-bound). Where performance results are not immediate and easily measured, organizational leaders as well as the Network leader must find creative means for motivating Network members and supporting extended members.

What Is the Level of Autonomy Allowed for the Network?

Networks should be allowed to operate with a fair amount of autonomy. Although the goals and objectives are fixed, the Network should be allowed to find creative, value maximization approaches to delivering on the goals and objectives. Essentially, Networks should be allowed to determine how to achieve their goals and objectives.

Too much oversight and control has the potential to limit and stifle creativity and solutions. On the other hand, too little control in the hands of immature and unseasoned leaders can result in inefficiencies and poor Network performance. The key to optimal Network performance resides in establishing the right balance of autonomy and control. Among the factors influencing this balance are the following:

1. Ambiguity around the goals and objectives
2. Leadership skills and capabilities of the Network leader
3. Organizational and leadership culture
4. Amounts of resource commitments required for operating the Network
5. Size of the prize and how urgently solutions are required

Is a Network Charter Available?

As discussed earlier, a charter helps to guide the Network in terms of key deliverables and a more defined SOW. In the absence of a defined charter with a clearly defined scope of work, the Network's activities can very easily veer away from priorities based on the interpretations and personal desires of the team. Furthermore, a more chaotic work environment may result as well-intended Network members extend their efforts to take on more than the available resources within the team can support. For Networks, the intention of a charter is to guide the team and keep it focused on deliverables. From a leadership perspective, be sure to provide at the very minimum, the strategic goals of the Network from which a more detailed work plan will evolve.

Getting Networks Started: Conferences, Training, and Chartering

Critical to the success of Networks, and in particular PSM Networks, is the need to have a consistent message regarding goals and objectives of the whole PSM initiative and how Networks fit into this bigger picture. In view of this and given that many organizations may consider setting up 12 to 14 Networks to support PSM work, a Network conference and workshop may be required.

A well-planned Network conference supported by senior leaders of the organization conveys the same message to all Network members in a consistent manner. When Networks are activated in a conference and workshop, with strong leadership support, confidence will generally resonate with all Network groups, creating an atmosphere of excitement and opportunities for all Network members.

Occasionally, other methods for activating Networks outside of a conference and workshop may arise. Activating them individually as the organizational needs arise is always a workable option. Care must be taken to ensure the Network requirements discussed earlier are properly met. Other methods used to activate Networks with varying levels of success may include the following:

- E-mail and other electronic media notification
- Direct leader notification and communication with individual members with Network leaders coordinating and activating the Network
- Video- and teleconferencing

Regardless of the method of activation, Networks must be activated properly for sustained success. The methodology for activating Networks in a conference and workshop is discussed next.

Defining the Need for a Network Conference and Workshop

Network conferences and workshops are most effective when there are large numbers of PSM Networks being activated at the same time. Activation via conferences and workshops is of even greater importance when the organization is very diverse, with several business units and membership of all Networks dispersed across all business units and across a wide geographical area. In such cases, the logistics of activating via conferences and workshops must be carefully planned and executed for success.

On the other hand, when the organization has few business units and members are located in the same geographical locale, planning and executing a Network conference and workshop may be logistically easier to do so. Network activation via conferences is essential when the following organizational conditions are prevailing:

1. Large numbers of Networks are being activated at the same time.
2. Network members are not very aware of their roles and responsibilities as a member of the Network.
3. There is a need to demonstrate leadership support and commitment to PSM and continuous improvements within the organization.

Planning and executing a Network conference and workshop is extremely complex and required to generate the desired results. Here, the authors seek to provide business leaders involved in PSM rollout and activation across their business the necessary requirements for a successful activation.

Objectives of Conferences and Workshops

Once all of the PSM Networks to be activated have been determined and all members of each Network have confirmed the right members into their respective roles, a Network conference and workshop can be used to activate these Network teams. The goals of the conference are generally as follows:

- The tactics to achieve the objectives and goals could be left to the subject matter experts who belong to the Networks. The subject matter experts in the Networks will have to provide workable solutions to challenges in the areas of focus of the Network.
- Define and confirm the role of a Network within the organization.
- Ensure a clear and common understanding of the goals of PSM Networks within the organization.
- Clarify responsibilities of Network members.
- Provide support and guidance to members regarding the ways Network members will fulfill their responsibilities to the Networks while they continue to perform the duties of a regular day job.
- Confirm and endorse resource commitment from senior leadership of the organization.
- Provide an opportunity for members to meet members of the same Network and begin the norming process in teams.
- Provide an opportunity for Network members to meet with and develop relationships with other Network members and subject matter experts.
- Provide face-to-face interactions with the senior leadership and Network sponsors.
- Develop an understanding of the magnitude of the undertaking for each Network and the level of commitment required from each team member.
- Provide an opportunity for Network members to learn new ways of doing work (change in thinking processes).

Against these objectives, conferences and workshop organizers will have to decide the best way to communicate the Network objectives and also achieve the aforementioned goals.

During the conference, senior leaders can outline the new direction and advantages of creating Networks to the organization. During the conference, all members hear the same message and have an opportunity to ask questions. Depending on the level of autonomy provided to Networks, leaders may communicate *what* the deliverables of a particular Network may be, while determining *how* these deliverables will be achieved is left up to the Network. Through a combination of well-planned conferences and workshops, the organization may create clear understanding of the deliverables of each Network and may initiate some of the planning processes in determining how these goals may be achieved. During the workshop, Network members can be challenged to think big (i.e., beyond their normal sphere of influence and control) and to focus on developing solutions that are beneficial to the entire organization. All members must work toward corporate solutions versus individual and departmental focus.

Current State versus Desired State Vision

When activating Networks, leadership should possess a clear understanding of the current state of its organization. There should be an understanding of the following:

- Where high-risk gaps may exist in the business processes.
- A list of the potential opportunities for improvement that may exist in the organization.
- Potential best practices that can be leveraged for corporate benefits.

Table 6.1 provides examples of current state versus desired state goals of Networks created by the organization. Careful analysis is required to identify and define the gaps between the current state and desired state of the organization. Gaps identified help to identify the types of Networks that are to be developed within the organization as well as the areas of focus of the various Networks.

Network Conference (Workshop) and Kickoff

Getting Networks off to a good start is an essential requirement for success. Remember, for many Network members, the work of the Network is to be undertaken in addition to their normal assigned work. In view of this, when activating Networks, members must be motivated, feel a sense of pride for being selected to function in Network roles, be recognized as subject matter experts, and be treated with respect and professionalism. Indeed, Network roles must be structured such that members of the general work population must see value in functioning in this capacity and should want to seek their turn at the opportunity to serve as a Network member.

A good way to kick off a group of Networks is to do so at a Network conference or workshop. In this way, all Networks are provided the same set of messages by senior leaders of the organization at the same time. More

TABLE 6.1

Current State Conditions versus End State Goals of the Organization

Current State	End (Desired) State
Pockets of excellence and best practices existing within the organization	Sharing of excellence and best practices across the entire organization
Lack of operational discipline and failure to consistently use procedures in achieving conformance to standards and policies	Consistent use and application of procedures to achieve conformance to standards and policies
Imbalances in distribution of skilled and experienced workers; employee capability and skills development	Subject matter experts identified and venues created for developing expertise and capabilities throughout the organization
History of major and catastrophic incidents	Creation of a learning organization; learning from incidents, understanding the root causes, and ensuring corrective actions are in place to address these root causes
Priorities change often	Focus on the corporate vision with aligned strategic goals
Weak leadership and management system	Development of an operationally excellent Management System
Inconsistent operational reliability	Safe and reliable operations and production systems
Leaders' commitment changes with changing work priorities	Leaders to follow through on their commitments; when commitments cannot be met, stakeholders are engaged to develop corrective measures
Lack of trust between leaders and employees	Trust developed through demonstrated leadership behaviors. Consistent application of the ABCD model for creating trust in the workplace (Blanchard, 2010)

important, at such a forum the various Network members get an opportunity to meet with the entire Network team and begin some of the face-to-face bonding that is so very important for cohesion in a dispersed team. This is particularly important for large and dispersed global organizations.

Pre-Network Conference (Workshop) Planning

Planning a Network conference is hard work and takes a lot of careful attention to get it right. Remember, the key is getting the Network off on the right footing. A poorly planned kick-off conference will result in low confidence and a less than optimally motivated Network team. Doing the hard work up front helps in setting all Networks up for success.

Preconference planning requires the conference planners to establish the following such that properly crafted messages for the conference are created. In view of this, leaders of the organization should be engaged to establish the following:

- Clear objectives and goals for each Network.
- Clearly defined gaps between the current state and the desired state for each Network and for the organization as a whole.
- The range of Networks required in supporting the organizational aspirations.
- The size of each Network based on the scope of work for closing identified gaps in each PSM element area.
- Skills, experience, and competency requirements for Network membership.
- Leadership support or sponsor for each Network.
- Resource and budgetary allocations for Network operations as well as for the Network conference.
- Criteria for Network success.
- Senior leadership involvement, commitment, and support for these Networks.
- Senior leadership presence during the duration of the Network conference.

When all the key steps are in place, it is the time to appoint a project sponsor and a project lead to start planning for the conference or workshop.

The processes and steps required to execute a Network conference in a manner to deliver the right levels of interest and commitment from senior leaders and Network members is explained in the following.

Planning and Executing the Network Conference

Planning a good Network conference requires several months of lead time (generally 6 to 8 months in advance of the event) to ensure the conference achieves its goals. Like any good project, a conference lead with a corporate sponsor is essential. Once the sponsor and the conference lead is in place, a conference supporting group or team is necessary to ensure the required work is completed consistent with conference preparation milestone targets. Support personnel with the following expertise are necessary:

- Office administration—For sending invitations to the Network members and for fielding questions that may arise from selected Network members. Also, this support group will assist in preparing and packaging required and essential conference materials for each Network member and conference participants.
- Finance—Discipline leads will work with the conference leader to ensure funds are properly sourced, allocated, and tracked.

- Communications—Subject matter experts in this discipline help in crafting the right messages and communication materials for presenters at the conference.
- Subject matter expert—PSM experts are required to help articulate the value of Networks for each PSM element as well as to ensure the accuracy and validity of PSM gaps that are identified and referenced in all communication materials.
- Supply chain management—Assists in procurement and logistics management. Generally, conferences of this type are best hosted outside of the organization so that work-related distractions are minimized.
- Project lead—The conference lead works with the team to ensure conference deliverables are met as per schedule. The project lead's primary role is to remove obstacles from the path of the working team so that conference deliverables in terms of participation, materials, Network member selection, travel, and accommodations are all met. As a result, participants arrive safely, are properly accommodated, and well equipped for an exciting PSM Network conference. Strong leadership and stakeholder management skills combined with good work ethics will only enhance the quality of conference and workshop activities.

An initial meeting that includes the sponsor in the conference planning team serves to clearly articulate the goals of the conference and to motivate the team to high levels of performance.

Once the conference planning team has been established, the steps and activities identified in Table 6.2 are essential for ensuring a successful Network conference.

TABLE 6.2

Steps and Activities Involved in Network Conference Planning

Step 1: Brainstorming	Step 2: Process Definition	Step 3: Resourcing
• Confirm conference objectives • Identify and agree on team responsibilities • Confirm the number of Networks to be established • Select conference venues • Secure resource and budget commitments • Agree on meeting frequency and norms	• Develop a conference planning charter • Identify and allocate tasks to achieve deliverables • Finalize conference objectives • Identify required attributes of conference venue and firm up agreement on dates, accommodation, and facilities required	• Issue meeting invitations for all Network members and conference participants • Develop conference presentation materials and key messages • Identify and remove obstacles as they become available

To support the leadership and outcome of the conference planning team, the following simple processes and tools may help the team lead with the work to be completed during this period.

- A simple one-page conference charter that clearly defines in/out of scope items.
- A simple action and decision log that lists decisions and actions including due dates and open/closed status.
- A risk log so that actions are prioritized based on the potential risk they may have on the outcome of the conference planning.

Conference Charter

To maintain focus on the conference deliverables, a conference charter helps in eliminating distractions and keeping the team on track. The charter should clearly describe the objectives, activities, tracking measures, and budget, in scope and out of scope items, and deliverables. A sample charter is provided in the subsequent chapter.

Action and Decision Log

The action and decision log document is a simple tracking tool required to ensure all required actions and decisions are tracked for keeping the project on schedule. This tracking tool must be accessible to all conference planning team members so that assigned tasks are not forgotten and delivered by the due dates required. Among the common actions required in planning the Network conference are as follows:

- Selecting the venue, meals, and venue logistics
- Listing and inviting Network members and participants
- Deciding on breakfast, lunch, and dinner menus
- Creating participant information package
- Determining the duration of the conference and its agenda
- Designing conference activity materials
- Selecting and training facilitators for individual Network activities during the conference
- Identifying keynote speakers (select someone with credibility in PSM Networks to lend credibility to the entire event)
- Confirming availability of key senior leaders (generally CEO or vice president of health, safety, security, and environment)

- Agreeing on and preparing key messages for senior leaders participating in the conference, including opening and closing remarks by key leaders
- Hiring public relations messaging and photographers to showcase event

Risk Log

The purpose of the risk log document is to support risk-based decision making by the conference planning team. Actions and deliverables that have a high risk potential to the delivery of a quality on budget and on schedule are identified and prioritized such that risks can be mitigated in an effective manner. Among the common risks that can impact the outcome of the Network conference are the following:

- Adequate backup plans for the venue in the event of venue-related problems
- Mitigation plans for bad weather and travel conditions
- Key speakers unable to attend and participate in the conference as planned
- Technological malfunctions (audio, video, communication equipment)
- Unplanned cost overruns
- Backup committee members

Leadership Commitment

Leaders at the most senior level must be committed to establishing and activating PSM Networks and must demonstrate their commitment by active participation in the conference and related activities. Strong engagement and involvement in preplanning work is required from senior leaders and they should be prepared to talk knowledgeably about the value and intent of the PSM Networks if approached by Network members during the conference.

Through regular involvement and engagement, senior leaders will be able to ensure consistency in key messages to Network members and to the organization as a whole. As should be the case in any strategic plan, the intent of senior leaders should clearly articulate the vision of the organization regarding PSM Networks. Key messages should highlight the desired state of the organization regarding PSM and PSM elements. The resounding message to the PSM Network members should be: *tell us how we get there and we shall remove the obstacles in your path to getting us there.*

Role of Senior Leaders during the Conference

During the conference, leaders should seek to create a shared PSM vision and clearly articulate the new direction of the organization. Among the key duties of senior leaders during the conference are:

- Be visible, engage participants in discussion regarding conference objectives and seek feedback.
- Be aligned in support and messaging for the value of the PSM Networks and leadership commitments.
- Create positive energy and inspire participants to share the vision regarding the desired process safety state.
- Attend conference activities where Network brainstorming may be planned.
- Understand and support participants' work.

Conference and Workshop Activities and Network Chartering

During the conference, planned Network activities should allow all Network members for each PSM element for which a Network is identified as essential to meet and become familiar with other Network team members. Network membership, in general, should be for no less than 2 years. With the support of the Network facilitators, conference activities should allow the Network to initiate work in developing an initial Network charter, which will likely guide the work of the Network for the medium term.

More important, this initial brainstorming period allows Network members to be comfortable with the time commitment required to fulfill the duties of a Network member. Furthermore, boundaries relating to what is in scope and out of scope for the Network begin to take shape. For example, in a Contractor Safety Management PSM Network, contractor relationship management may be out of scope, whereas contractor prequalification will fall within scope of the Network. The Network participants must develop a strategy and tactics to bring the shared vision to reality.

At a conference Network members should:

- Obtain a clear understanding of the purpose of the Networks.
- Recognize the benefits to the organization from the work to be performed by the Networks.
- Develop and validate a Network charter that should include but not be limited to:
  - Clearly defined measurable goals and objectives of the Network
  - Key milestones and deliverables
  - Action management and tracking process

    - Defining the scope of work (what is in scope and what is out of scope for the Network)
    - Identify key stakeholder groups for which RACI obligations are identified
    - Confirm that Network members share the organizational vision and are the right appointees for membership
    - Type of Network engagement, type and frequency of meetings
    - Metrics for measuring success
    - Frequency of updates to senior leadership regarding Network activities
- Development of an initial Network work plan:
    - Brainstorm on the work activities required to deliver on the identified deliverables of the Network charter
    - Prioritize work activities identified
    - Select and prioritize big-win opportunities for the organization
    - Identify and use simple tools and processes for generating Network value, for example:
        - Conduct a gap assessment between current and desired state to identify big-win opportunities for the organization
- Develop a prioritized action–decision register complete with assigned responsibilities of team members and due dates for closing prioritized gaps

Key Messages for the Conference

The importance of having the right messages to support Network activation during the conference cannot be understated. This is perhaps the best moment for confirming to all Network members that the organization is committed to improving PSM performance. In view of this, the following key messages should resonate among Network members:

- A clearly articulated PSM vision is shared.
- Organizational and leadership commitment to improving PSM.
- Through the efforts of the Networks, the organization will be brought closer to its shared vision.
- Networks will be properly resourced and supported by leadership.
- Several Networks may overlap in scope; seek out these overlaps and mange accordingly to avoid duplication while at the same time optimizing resources.

- Note measured progress and key performance indicators (KPIs) essential for gauging progress and success.
- Focus on the big-win opportunities, sweat the small stuff later.

Reference

Inkpen, A. C., and Tsang, E. W. K. (2005). Social capital, networks and knowledge transfer. *Academy of Management Review,* 30(1): 146–165. Retrieved April 12, 2012, from EBSCOhost database.

7

Activation and Tenure of Networks

Successful Networks rely upon proper activation, management, and support. Networks provide a huge opportunity for leveraging a company's subject matter expertise and knowledge in the continuous improvement of its business. This chapter answers the following questions:

- Why is it necessary to formally activate Networks?
- Why are control of membership and membership changes necessary?
- What value is there in sustaining Network activities?

Networks: How They Differ from Other Organizational Structures

Networks are organized differently from traditional organizational structures such as departments, work groups, and project teams. The principal functions of Networks are to capture and exchange knowledge, share learnings, and build organizational competency in the specific areas of focus of the Network. Table 7.1 provides a comparison of various organizational groups and their functionalities.

What differentiates Networks from other teams and committees is the unmistakable passion for the discipline topic and area of focus for continuous improvements that the Network is focused on. Where Process Safety Management (PSM) is concerned, Network members usually share a very strong mission to share knowledge and learnings in an effort to prevent the occurrence of high consequence, low probability events. Network membership passion may be the result of very personal experiences in PSM or may be the outcome of a genuine love of science and engineering. Regardless of the source of the passion, this attribute is a strong success factor in the effectiveness and sustainability of Networks.

TABLE 7.1
Comparison of Organizational Groups

	What Is Its Purpose?	Who Belongs to It?	What Holds It Together?	How Long Does It Last?
Network (or centers of excellence)	Build and exchange knowledge (business focus) in the area of focus of the Network	Members are usually, but not always, self-selected or join voluntarily; some organizations nominate members based on experience and qualification	Passion and commitment around day jobs and perceived value contribution opportunities for the organization	As long as there is business value creation, opportunities to continue to improve and Network passion
Organization/department/function	Deliver a specific product or service to the overall organization	Often automatic or hired into on a permanent basis	Common goals and job requirements	Until the next reorganization (hierarchical) occurs in the business or promotion
Work group or project team	Accomplish a specific task deemed essential by the organization	Determined by management assignment	Project milestones and other project constraints such as budgets	Disperses when the project is complete
Committee	Consider, investigate, or report on a matter that arises	Delegated or self-nominated	Sense of common purpose or goal	Stays in place until decisions are made (temporary); transient and very focused

Source: Ranta, D., 2011, APQC case study: Engagement and participation for knowledge sharing and collaboration, retrieved October 25, 2012, from www.apqc.org.

Why Is It Necessary to Formally Activate Networks?

Recognizing that Networks typically function across many organizational boundaries, it is also important that they be sanctioned with appropriate mandates by leaders or sponsors who have the authority and responsibility to assign priorities and resources. The need for starting a Network begins with the identification of a business rationale. Formally sanctioning Networks requires leadership support and generally includes:

- Defined charters with clear deliverables and measures of success
- Stewardship to sponsors or a steering committee on a regular basis
- Description of Network member roles and skill requirements
- Visibility of Network activity through effective communication

There are five key steps in sanctioning a new Network in an organization, as illustrated in Figure 7.1. Following confirmation of the business need, it is critical that stakeholders are aligned on the Network work scope, objectives, and deliverables, which are embedded in the charter. With business need confirmed and the draft charter developed, the corporate approval process

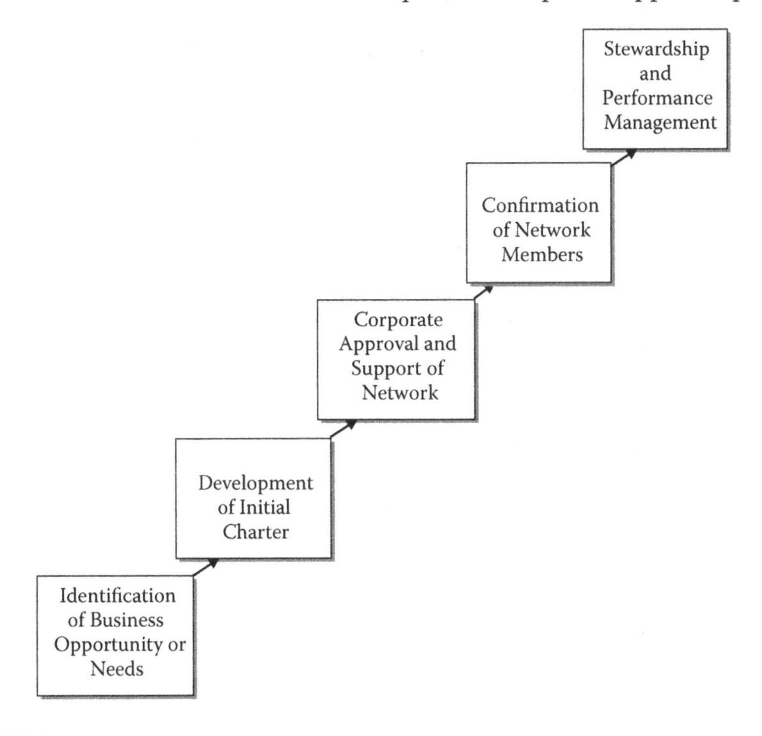

FIGURE 7.1
Steps in the activation of a Network.

sanctions the formation of a Network, thereby authorizing the allocation of resources to form the Network and for the Network to perform its duties.

In the case of Networks driving consistent compliance to PSM standards and application of best practices, Network composition and membership must be carefully considered to ensure the right subject matter expertise is embedded in the Network. The Network leader has the responsibility to facilitate knowledge transfer between the members, capturing best practices and sharing learnings. The Network leader needs to have a deep knowledge of the PSM element that the Network has accountability to support.

Why Are Control of Membership and Membership Changes Necessary?

Network formation is quite costly and requires extensive planning and organization. In some instances, the duties of Network members form part of the annual work plans of the member. In other instances, some members may be allocated fully to Network activities throughout the tenure of the Network. Network members also should have some subject matter expertise so that they contribute to Network discussions and activity in a meaningful way bringing field experience and technical competence to an already existing group of similar knowledge and expertise.

Success of Networks depends heavily on the credibility of the Network members. This is particularly so when stakeholder representation is required. Members should be strong respected leaders known for their impartiality and drive toward business improvements. Members should be seen as natural go-to people for the subject and be able to influence stakeholders to make change happen at the stakeholder group level.

Network members are expected to communicate Network activities and recommended gap closure work to local site stakeholders and the Community of Practice to ensure awareness, understanding, and commitment. Network members also are expected to bring site input back to the Core Team to vet issues and concerns. These responsibilities reinforce the necessary competence when selecting among the many potential Network members within the organization.

What Value Is There in Sustaining Network Activities?

There is an inherently strong business rationale for employing Networks in most organizations. No one person in an organization knows what the entire organization knows. The more an organization links individuals with subject matter expertise, the better chance the organization has to tap into pockets of best practices and learnings that can benefit the whole company. Networks enable access to knowledge and continuous improvements where organizational and hierarchical structure or functional barriers provide obstacles that hinder the flow of information and knowledge throughout the organization.

Generally, expertise and pockets of best practice are dispersed in a company. Very rarely do companies centralize all of their technical or subject matter expertise. Enabling an organization to utilize more of its inherent expertise and tacit knowledge when and where it is required unleashes the value maximizing potential of the organization. Benefits derived from Networks increase as the organization size and number of operating sites increases.

Because many organizations suffer from the silo phenomenon, they:

- Often experience the deficiencies and inefficiencies, and repeat the same incident at multiple facilities.
- Often solve common problems multiple times independently using scarce resources that could have been otherwise better used.
- Often implement standards and practices in different ways with different tools and processes, thereby limiting standardization opportunities and failing to learn from each other.

Each Network should have a specific business case based on PSM, Management System discipline, or topical area the Network is focusing on. Business factors generally considered include, but are not limited to:

- Risk reduction
- Value maximization opportunities, impact on bottom line
- Level of commonality and standardization required

Network Tenure

Once sanctioned and operating for a period of time, Networks will have to deal with changes in membership due to normal employee attrition and organizational change. If sanctioned, chartered, and stewarded well, most technically oriented Networks will become integrated in the organization. This implies that as long as the organization continues to require that type of technical competency to support safe and reliable operations, chances are the Network will be around a long time.

Networks will eventually become part of the organizational language, culture, and collaboration processes. In addition, participation and learning from the Networks gets institutionalized in development and progression plans. In order for Networks to have some stability and to ensure adequate development of members, it is recommended that Network assignments be a minimum of 2 to 3 years. This minimum period is required since it provides adequate time for members to move through the various phases of team building from storming, through norming, into collaborative work and production.

Network Membership Changes and Turnover

As with any organizational role, people move from one role to another for a variety of reasons. Similarly, members of the Network will consider opportunities when they arise and from time to time Network membership will change. Because Network roles are deemed to be critical positions and personnel are intensively screened to determine eligibility for membership, it is recommended that membership changes follow a rigorous management of change process. In this way, risk can properly be assessed and appropriate replacement candidates can be identified and oriented to the role.

The rigor associated with managing the change should be based on the level of risk of the change. For example, consideration will be given to the following:

- Is the Network focused on process safety critical work?
- Are the competencies of the Network members very important for the success of the business?

When a Network leader or Network participant moves to a new position, their Network role may or may not shift or change. It is possible for someone to have multiple organizational line roles and still be a participating member of a Network. Ideally, Network members should be asked to commit to the role for the minimum 2- to 3-year tenure as suggested.

When a change in role occurs, care should be taken to ensure an equivalent subject matter expert (SME) and capable replacement is found. Here, support from senior leadership is essential to navigate challenges and obstacles to finding a suitable stakeholder representative. Timely membership replacement helps in keeping the Network focus steadfast and in meeting scheduled delivery of workable business solutions to the organization.

In addition to issues of turnover, member contribution and support can be a constraint on the effectiveness of a Network. The Network leader should be vigilant on member contributions, tardiness, absenteeism, and other factors that contribute to poor performance. The details behind a member's absence must be investigated, and action needs to be taken to ensure that the correct business value is being delivered. If someone, for whatever reason, needs to have their position replaced, the Network leader—with support from the Network stewardship organization—should initiate the necessary actions to identify and select a suitable replacement. Both membership turnover and lack of participation by members may result in reduced Network effectiveness. Progress on deliverables is slowed, time is wasted reorienting members on missed topics, and morale of the remaining Network members can be negatively affected. Leaders need to support stability of the Networks in order to avoid lost Network productivity and morale. The importance of leadership in supporting effective Networks is discussed in more detail in the subsequent chapters.

Reference

Ranta, D. (2011). APQC case study: Engagement and participation for knowledge sharing and collaboration. Retrieved October 25, 2012, from www.apqc.org.

8

Network Focus and Work Priorities

Within an organization, Networks are developed to generate continuous improvements in business practices or in core business areas of the organization. Networks commonly found in organizations may include the following:

1. Management Systems Networks
2. Process Safety Management (PSM) Networks
3. Specialty engineering area Networks, such as electrical, rotating equipment, and instrumentation Networks

The goals of Networks are collaboration across geographical, technological, and experience boundaries to find creative solutions to complex problems faced by organizations in an organized manner. When Networks are activated in organizations, they are done so purposefully and with great care and attention to address value-added opportunities identified by the organizations and which require focused attention by a group of people working together for a common cause.

Today's highly competitive business environments require that organizations continue to evolve and to find creative means for generating continuous improvements within the business environments. The creation of Networks helps organizations remove obstacles and develop new ways for doing work better, faster, and in a more cost-effective way. Real value is generated from Networks when they are structured, supported, and provided clear goals and deliverables.

Types of Networks Developed in Organizations

Organized and formal Networks in organizations are fairly new concepts that are now being leveraged to generate continuous improvements in organizations. Networks are identified based on the following:

1. Gaps in organization between the current state and desired state of the organization.
2. Opportunities in the business environment that require an organized team effort of subject matter experts (SMEs).

3. Persistent problems in the operations and Management Systems of the organization.
4. Rolling out of new standards, policies, or even Management Systems and initiatives.

Networks today, however, fall into the three main categories: Management Systems Networks, PSM Networks, and Specialty Engineering Area Networks. Furthermore, some Networks may be industry specific, whereas others may be generic and can apply to any industry.

Organizations must apply caution when establishing Networks to ensure areas of overlap lead to the forming of a single Network as opposed to multiple Networks that may be required by both Management Systems and PSM, for example. This is clearly demonstrated in the instance where the organization may seek to establish a Contractor Management Network, whereas PSM may seek to establish a Contractor Safety Network. In such instances, where areas of overlap are so great, there is tremendous value in establishing a single Contractor Management Network at the management system level depending on the level of maturity of the organization at the PSM level. Where PSM maturity is low, it may be best to start with a PSM Network that evolves into a Management System Network as the maturity of the PSM Network grows and business performance improves to sustainable levels.

Table 8.1 provides a list of potential Networks that organizations may activate to address and generate continuous improvements in the business. The focus of Networks moves from technical through strategic from operations management through PSM to Management Systems Networks.

Network Focus

The primary focus of all Networks in organizations is to achieve the following:

1. Reduce risk exposures of the organization
2. Improve processes, procedures, and work practices
3. Solve technical problems and concerns in the organization
4. Improve operating efficiency
5. Improve the strategic position of the organization
6. Create value and improve value maximization in the organization
7. Increase organizational competency

To keep Network teams focused on these goals, a structured approach to retaining Network attention is required.

TABLE 8.1

Common Networks Generally Activated by Organizations

Operation Management	PSM	Management System
1. Hydrogen production 2. Hydroprocessing 3. Naphtha reformer 4. Hydrocracking 5. Sulfur and amine 6. Maintenance planning and scheduling 7. Water treatment 8. Refinery turnaround 9. Operations management 10. Process and automation 11. Electrical 12. Mechanical 13. Rotating equipment 14. Engineers in training	1. Emergency management and preparedness 2. Learning and competency assurance 3. Contractor safety 4. Incident investigation 5. Quality assurance 6. Management of change 7. Audit and inspections 8. Risk management	1. Leadership, management, and organizational commitment 2. Security management and emergency preparedness 3. Qualification, orientation, training, and competency 4. Contractor/supplier management 5. Event management and learning 6. Management of engineered, nonengineered, and people change 7. Risk management 8. Audits and assessments 9. Documentation management and process safety information 10. Process hazard analysis (PHA) 11. Physical asset system integrity, reliability, and quality assurance 12. Operating procedure and safe work practices 13. Legal and regulatory

As discussed earlier in Chapter 5, Networks are generally comprised of a Core Team comprised of six to eight carefully selected workers; an available and accessible pool of SMEs; and a wider, more distributed, and larger Community of Practice. Members of the core Network team, in addition to their day-to-day duties, are required to solve complex problems and generate value for the organization. This group taps into the SMEs (or extended Network) for technical expertise to do so. Once solutions are generated, the SMEs and Communities of Practice execute and activate at the frontline.

To do this in an organized way, the Network team must work cohesively together and should not be distracted by outlier activities or out of scope items. The Core Team should be guided by a Network charter that identifies the big-win opportunities identified by the organization and by the Core Team members for achieving the goals identified earlier. These big-win opportunities are generally reflected in risk exposure reduction, cost and

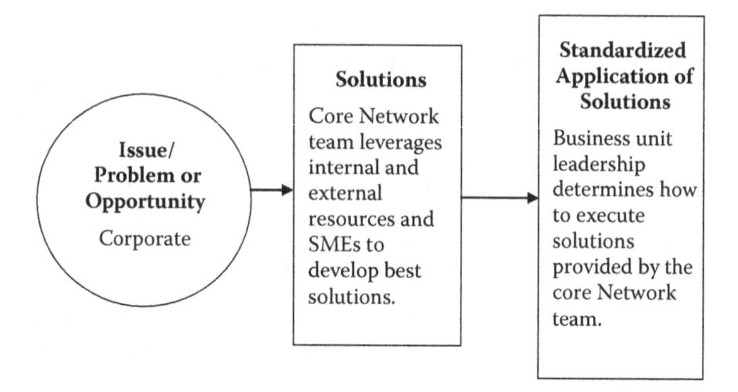

FIGURE 8.1
Network approach to value creation in the organization.

efficiency improvements, fewer unplanned shutdowns, and improvements in overall operational discipline.

More important, when solutions are developed, they are shared across the organization and executed in a standardized way as best possible across the entire organization. In many instances, however, Networks will determine *what is required* to address the business opportunities or needs, while the business unit leadership will often address *how the solution will be applied* across the business unit. Figure 8.1 demonstrates how the Network functions.

Network Charter

In developing its work plan and work priorities, a Network must start first with a Network charter. The Network charter is a useful tool to keep the Network focused on its work priorities. What should a Network charter entail? A well-written charter will guide the Network activities, and provide the members a simple and consistent road map. The charter will also keep Network members focused on the work scope and key deliverables. In the absence of a clearly defined and approved charter, the Network may gravitate from the original scope of work toward preferred areas of interest. Network charter should include the following:

- Measurable goals and objectives
- Critical success factors
- In scope and out of scope items
- Clearly defined deliverables
- Team members

- Meetings frequency and norms
- Leadership sponsor and approval

Table 8.2 provides an overview of a simplified one-page charter for a typical Network. A simple charter makes reference to this tool as a simple and uncomplicated exercise.

Measurable Objectives and Goals

When Networks are set up, senior leadership of the organization may have strategic goals and objectives identified for the Network. To enhance the functioning of the Network, members must translate these strategic objectives into measurable goals and objectives. These more finite goals and objectives provide opportunities for the Network to monitor its progress and overall performance toward achieving the strategic objectives of the organization. Where applicable, goals should follow the SMART principle (specific, measurable, achievable, realistic, and time-bound).

Critical Success Factors

For the Network, an objective assessment of what success looks like is required. Members should establish measures to determine the success of the Network. Ultimately, the goal of Networks is to create and maximize value for the organization. The Network should also identify risk factors that may prevent the Network from achieving its goals and with the help of its senior leader sponsor, work to mitigate these risks to manageable and acceptable levels.

TABLE 8.2

Sample One-Page Network Charter

Measurable Goals and Objectives	**Critical Success Factors**	**In-Scope**	**Out of Scope**
Deliverables	**Meetings Frequency and Norms**	**Team Members and Tenure**	**Leadership Sponsor and Approval**

In Scope and Out of Scope Items

In scope and out of scope items is one of the key sections of the charter document; the Network team must decide what team activities are in scope. In scope items establish manageable boundaries for the Network. Business environment changes that may influence the operations of the business may also impact in scope and out of scope items for the Network. When determining the scope of work for the Network, members should be realistic in determining what can be reasonably accomplished based on resource constraints imposed upon the Network. Listing out of scope items is also a good way to ensure the Network remains focused. Out of scope items may be those dealt with directly by the business unit or perhaps another Network. The intent of delineating out of scope items is to avoid costly duplication and to ensure the Network focus on big-win opportunities.

Clearly Defined Deliverables

In a charter it is important to identify the deliverables of the Network. These are more granular than the goals and objectives of the Network and similarly follow the SMART principle. Deliverables can be broken further into specific work packets that can be included in the work plans of each Network. Deliverables may include items such as training material, communication plan, and change management strategy.

Team Members

When completing the charter, the Network must identify all team members and, where possible, stakeholder interest representation. A key consideration is ensuring adequate representation from all areas of an organization. Team member selection criteria must be established and members should be knowledgeable on the areas of focus of the Network. Assigning junior employees who do not have in depth and breadth of knowledge about the subject will place the Network at a disadvantage in meeting its deliverables.

Members of the Network should be equally competent as best possible. Such balances in competence helps the team to continue to function in the absence of one or two of its members. Members must be skilled leaders and workers who can work independently on tight schedules, able to research, have well-developed industry contacts and informal Networks, excellent communication skills, team skills, and nonauthoritative leadership skills. *Network members must be respected employees within the organization at middle to upper management levels whose authority is derived from their knowledge and expertise in the areas of focus of the Network.*

Meeting Frequencies

Meeting frequencies depends upon the type of work being performed by the Network and the maturity of the organization in the area of focus of the Network. For organizations that are relatively matured in the areas of focus of the Network, meetings will be less frequent than those organizations that are immature. Typically, Networks will meet biweekly in the initial stages of kickoff and will transition to monthly meetings once a work plan has been developed and members are assigned specific tasks and work packages.

Cost management is an essential requirement for Networks. For a distributed team, leveraging technology in the areas of teleconferencing, live meeting, Lync, SharePoint, collaboration sites, and other virtual technology serve as an effective means for conducting meetings while managing costs. However, there is nothing more powerful than face-to-face meetings when resolving complex issues. Networks must, therefore, establish the right balance between virtual and face-to-face meetings for meeting charter obligations.

Leadership Sponsor and Approval

Organizational and senior leadership approval in many instances is required since resources commitment is required to achieve the goals defined in the charter. As a consequence, a senior leadership sponsor is necessary to support the Network activities. In this section, members of the Network must also affix their commitment to the goals and objectives and other deliverables and conditions identified in the charter.

Also in this section, it is generally useful to commit members and leadership to a finite period. By doing so, turnover among members is limited, and members are generally more committed and motivated to provide solutions before their tenure expires. Generally, tenure for Network members ranges from 2 to 3 years. Membership to the Network must be something other members of the workforce aspire for as a coveted role. Furthermore, to avoid burnout and to provide development opportunities for other members of the workforce, it is recommended that Network membership be tenured.

Network Work Plan

To generate best performance from the Network, it is recommended that the Network develop an annual work plan. Typical activities contained in the work plan may include the following:

- Review of the organization's policy, values, standards or procedures, and work practices related to the Network focus and identify required improvements to achieve the desired state for the organization.
- Review the existing systems and processes including tools, templates, and technology for supporting the desired state requirements.
- Develop stakeholder engagement strategy.
- Develop required communication and change management strategy for impacted stakeholders.
- Develop required training materials.

The work plan is a simple tool that translates each deliverable identified in the charter into manageable work packages such that the deliverable can be achieved. The work identifies what is required and how the members shall achieve it. Figure 8.2 demonstrates a simplified version of the work plan process for achieving the charter deliverables.

When developing the work plan, the Network should take time to ensure the workplace deliverables as well as the work plan activities are aligned with the corporate vision. When a Network is activated, achieving the vision of the organization may be supported by several deliverables. Furthermore, to meet the deliverable requirements, there may be several work plan activities

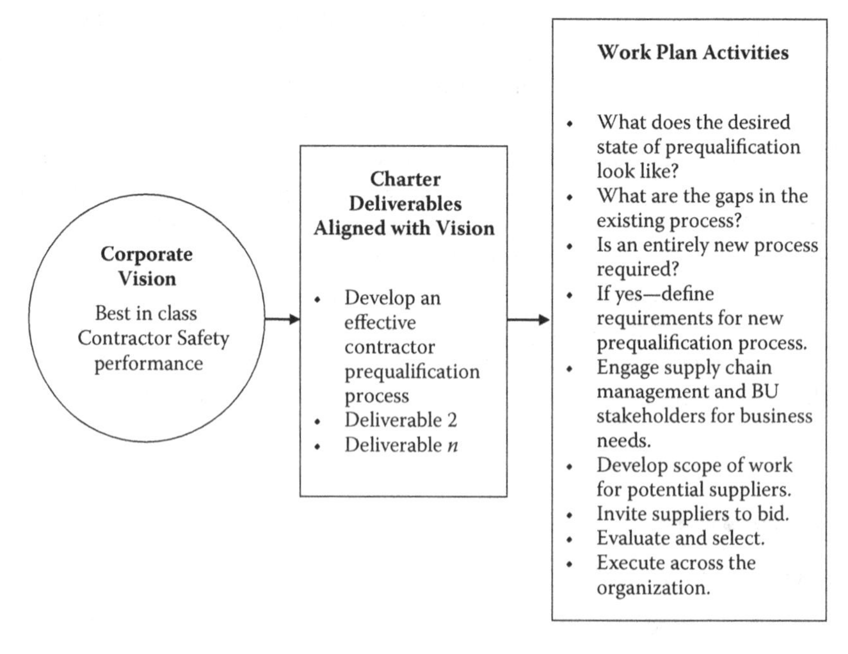

FIGURE 8.2
Demonstration of work plan activities for achieving deliverables.

that are necessary to achieve this deliverable. As a consequence, the Network must therefore apply prudence and appropriately prioritize work in order to achieve its deliverables.

Network Work Prioritization

Many tools and processes are available to assist workers in prioritizing work. In most organizations, work is prioritized based on the impact on production. Other organizations have evolved processes to prioritize work based on the following hierarchy:

- If the existing condition poses imminent danger to personnel, the environment, communities, assets and facilities, and company image and goodwill in that order.
- If the existing condition has the potential to impact production.
- If the proposed work has the potential to improve production, efficiency, and performance.
- If the work being performed is a need to have or a nice to have requirement.

A very effective method used for prioritizing the work of the Network is opportunity matrix. The *opportunity matrix* helps businesses prioritize work based on the *value creation* and *complexity in execution* relationship. Figure 8.3 demonstrates the use of this relationship to prioritize the work of the Network. In this example, prioritization is a reflection of work deliverables associated with a Contractor Safety Management Network.

In the example provided, numeric assessments of both complexity and value creation from each charter deliverable are determined by the Network. As shown in the matrix, deliverables that score closer to zero for both criteria are considered higher priority deliverables. Such deliverables are high in value creation and low in complexity to execute. At the other extreme, deliverables that score high for both criteria are low in value creation and complex to execute, and are therefore the lowest priority for execution.

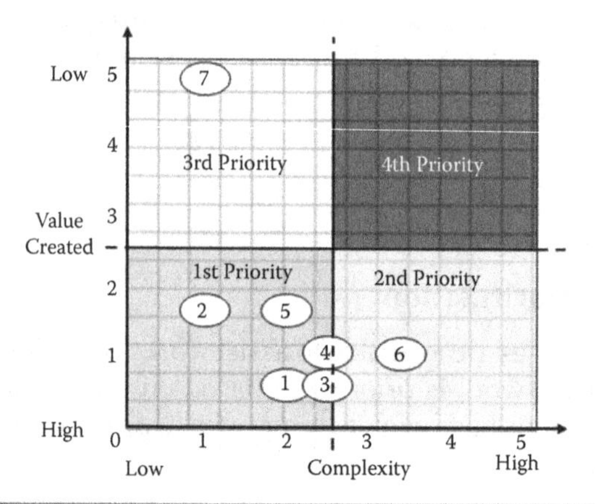

Priority	Opportunity	Complexity 1=Low 5=High	Value Created 1=High 5=Low
1	Get prequalification rolled out right	2.0	0.5
2	Align contract language with PSM/OMS requirements	1.0	1.5
3	Standardized Computer-Based Contractor Orientation (3 levels)	2.5	0.5
4	Simplify CM101 and make BU/BA specific (workshop with tools)	2.5	1.0
5	CM101 support to BU in rollout of Contractor Safety	2.0	2.0
6	Provide access to organization's work related information (procedures, policies, processes) to all contractors	3.5	1.0
7	Update the Contractor Safety Standard based on improvement opportunities identified	1.0	5.0

FIGURE 8.3
Opportunity matrix process for prioritizing work for Networks. (From https://www.safetyerudite.com/. © 2012 by Safety Erudite Inc. With permission.)

9

Establishing Key Performance Indicators (KPIs) for Stewarding Organizational Performance

A performance indicator or key performance indicator (KPI) is industry jargon for a type of business performance measurement. KPIs are commonly used by an organization to evaluate its success or the success of a particular activity in which it is engaged. Sometimes success is defined in terms of making progress toward strategic goals; but often, success is simply the repeated achievement of some level of operational goal, such as zero incidents or fatalities in the workplace. Choosing the right KPIs, therefore, is reliant upon having a good understanding of what is important to the organization.

Selecting, developing, and establishing good performance indicators are challenging and difficult exercises for the organization. Resources are committed in such exercises and must therefore be treated as any other business investment from a value maximization perspective. There are many things to measure on a process plant, but there are potential pitfalls in gathering too much data on too many metrics. The burden on the company's workforce could become disproportionate to their other day-to-day tasks to ensure safe and reliable operations. Networks are therefore an excellent resource to the organization in selecting, developing, and establishing credible and value maximization performance indicators.

The resultant plethora of data may also be complex to analyze and the key trends from the most insightful metrics may be obscured. Thus, care must be taken to regularly review the effectiveness of even well-established industry metrics to ensure that they remain relevant and insightful; otherwise, their use may detract attention away from the most material and important risks (Broadribb et al., 2010). A more holistic approach to performance assessment is required based on leading indicators and lagging indicators. Leading indicators measure parameters that are proactive and help prevent accidents from happening when timely intervention occurs. Lagging indicators measure the outcome of an incident or situation after it has happened (Fishwick, 2010).

Once Networks have been established and activated across the organization, each Network is required to identify and establish KPIs to help the business focus on its priorities that are aligned with the corporate vision. KPIs identified must be of high value and must meet the SMART criteria (specific, measurable, achievable, realistic, and time-bound) for effective stewardship. The goal of KPIs is to help guide business performance on an ongoing basis.

Networks are also effective in establishing the requirements for reporting on KPIs to ensure continuous improvements. Reporting frequency is intended to bring value to the organization by effecting required corrective actions on a timely basis. Furthermore, reporting frequency should be such that it does not tax the business unnecessarily in resource requirements and commitments for data collection and management related to the particular KPI.

Performance Targets

The term *target* tends to be used quite loosely alongside the related (although arguably distinct) concepts of *mission, vision, value, aspiration, aim, goal,* and *objective.* KPIs, however, must be distinguished from aspirational goals (e.g., zero accidents) as they are usually more tangible and realistically achievable.

When applied properly, KPIs can impact effectiveness in the following areas:

- Inputs—People and raw materials, processes and systems, and facilities and technology.
- Outputs—Activities and outcomes or results.

In practice, measuring inputs, outputs, and outcomes effectively poses a number of practical challenges in the business processes. Methods of measurements may include the following:

- Continuous or whole population monitoring—Can measure accidents, routine health surveillance.
- Sampling techniques—Can measure health and Safety Management System audits, safety climate surveys, behavioral surveys.
- Routine monitoring processes—Can measure planned or periodic inspections, and monitoring levels of workplace contaminants.

Regardless of the methods of measurement considered, there are likely to be many factors affecting the practicability and efficacy of measurement in each case.

In large organizations, some quantitative and qualitative targets may be the outcomes of a best guess based on professional judgment. At the other extreme, targets may be the outcomes of detailed scientific studies aimed at improving the organization's performance over an extended period. Ultimately, the goal is to establish meaningful targets that create a fire in the hearts of all workers to aspire for, encouraging them to perform at higher levels of capabilities in order to see the organization achieve stronger performance relative to the target established.

Meaningful target setting needs to be underpinned by a robust understanding of current performance status, including continuing problems, and their causes and possible solutions. Target setting in the area of health and safety needs to be understood in the context of an organization's approach to target setting in general; and in turn, this needs to be viewed in relation to its overall approach to strategic decision making. It is of the view that targets are most likely to be effective in helping to leverage change when they are evidence based and they are set with the active involvement of stakeholder groups or individuals who will be held accountable for delivering against them.

Performance measurement is of increasing interest to managers due to the changing nature of work, increasing competition, specific improvement initiatives like Six Sigma and strategic planning, national and international awards, changing organizational roles, changing external demands, and the power of information technology (Brooks and Coleman, 2003). Performance targets are the *success measures* of the organization's performance management system and are defined by KPIs. Without performance targets, the organization's vision cannot be quantified. A performance target provides an unambiguous definition of success. It signals what is important and it tells people what is expected. A performance target is assigned to a measurable strategy map item and run over a specific period of time and defined by the performance indicator it is assigned to.

Key Performance Indicators (KPIs)

Performance indicators are generally linked to a performance target. KPIs provide the definition around a performance target and are intended to help gauge the organization's management system in the area of focus of the KPI for progress and continuous improvements. As shown in Figure 9.1, each element of Process Safety Management (PSM) or the Management System

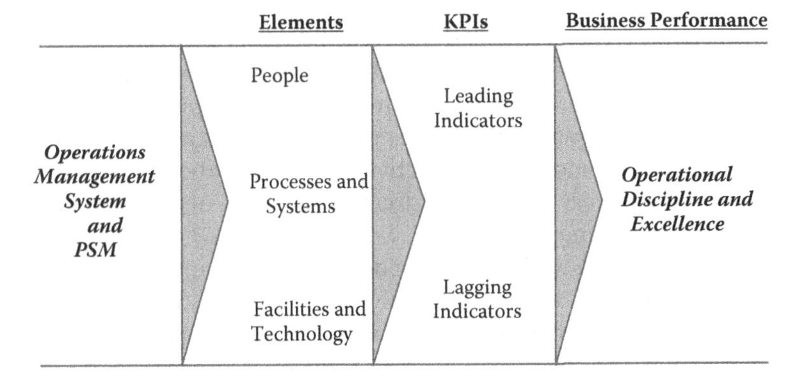

FIGURE 9.1
Simple model for establishing KPIs. (From https://www.safetyerudite.com/. © 2012 by Safety Erudite Inc. With permission.)

of the organization can yield *leading* or *lagging* indicators for each of which a KPI can be generated.

KPI is not a new concept; however, historical approaches to developing KPIs in organizations have been limited by a failure to appreciate the value of KPIs and selection of KPIs that were generally not aligned with high-value opportunities for the organization. Furthermore, leaders selected KPIs that were easy to monitor and for which data was readily available. Networks, when activated, have the ability to work with stakeholder groups to identify and prioritize high-value KPIs for organizational stewardship.

Working with the subject matter experts (SMEs) and members of the Communities of Practice, Networks are able to determine those KPIs associated with each element. Working together, the Network is also able to determine the following:

- Right reporting frequencies
- How best to collect the required data
- Methods of calculation and quantification of the KPI
- Which stakeholder group may be best able to collect the required data
- The value implication of the KPI for the organization
- Report on value generated to the business from stewardship of the KPI

Figure 9.2 provides a more detailed overview of the process for establishing and stewarding KPIs within the organization.

Leading Indicators

Leading indicators are proactive measures of the performance of key work processes or systems against relevant internal standards. Leading indicators aim at finding problems before incidents or near misses occur. A leading indicator generally provides assurance that the process is operating within specified performance standards. Leading indicators are considered the *drivers* of lagging indicators. What this means is that improved performance in a leading indicator will drive better performance in a lagging indicator. When measured and monitored effectively, leading indicators provide data to enable effective intervention to address or reverse a negative trend before it results in injury, damage, or loss.

Following the Texas City Refinery accident in 2005, an internal investigation identified that although the site had numerous measures for tracking operational and safety performance, these measures did not focus on leading indicators that would provide early warning of potential major incidents. Similar recommendations were made by the Baker Panel in its report on the safety culture in BP's five U.S. refineries. The independent panel urged BP to develop and implement process safety performance indicators (PSPIs) and then subsequently

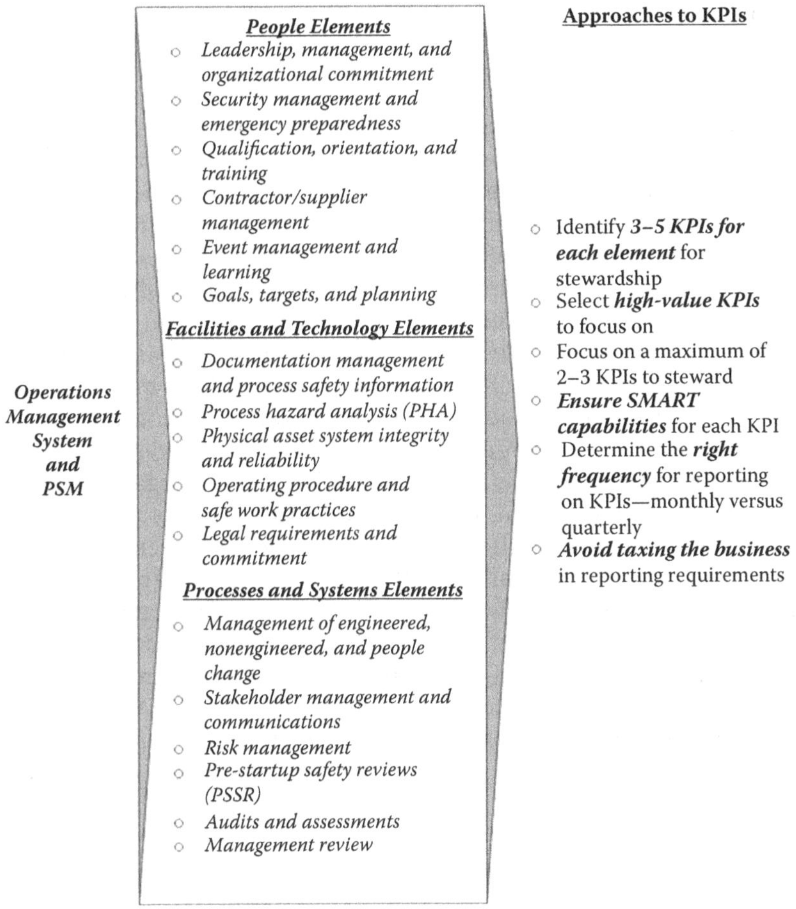

FIGURE 9.2
More detailed process for establishing and stewarding KPIs. (From https://www.safetyerudite.com/. © 2012 by Safety Erudite Inc. With permission.)

work within the industry to encourage their further development and adoption. This has resulted in BP substantially strengthening its process safety risk management through a comprehensive system of controls embedded within its Operating Management System (OMS), which is being implemented at the site level across all of its operations globally, and which are supported by good PSPIs to measure, evaluate, and correct performance (Broadribb et al., 2010).

Leading metrics look toward the future and are proactive. They indicate the performance of work processes, operating discipline, or layers of protection designed to prevent incidents. The less robust these preventative processes are, the more likely a threshold incident will occur. Both lagging and leading indicators are seen as critical to driving continuous improvement in process safety.

Lagging Indicators

Lagging indicators are reactive measures of some aspect of a process that has failed. These failures can have little or no consequences (such as a near miss) or can have large consequences (such as catastrophic failures) resulting in a loss of containment, fires and explosions, fatalities, and damage to the reputation of the organization. The common aspect of these events is that a layer of protection of a process safety system has actually failed. In many of these events, leading indicators, such as activation of relief valves, critical alarms, and minor leaks and spills, means that at least one layer of protection has failed and been largely ignored prior to the event.

For lagging indicators the level of investigation into the causes of the event will vary. The investigation into a fire could be different to a minor spill. Even a minor spill is important enough to be investigated and recorded. Trends for near misses should be recorded. An example could be several minor spills from one operating area. This may be an indication of a larger systemic failure in a maintenance system, which could in turn lead to a major process safety event. If these minor issues are not recorded and trends considered, then an emerging systemic failure pattern may not be recognized.

Leading indicators are differentiated from lagging indicators since leading indicators are regarded as performance drivers, whereas lagging indicators reflect outcomes of failed or ignored protection measures (Plastics and Chemicals Industries Association [PACIA], 2010). Lagging metrics are based on incidents that have already occurred. These incidents meet a threshold of severity that should be reported as part of the industry-wide process safety metric.

Typical Process Safety Management (PSM)/ Management System KPIs Generated by Networks

As discussed earlier, leading indicators are proactive and are significant drivers of strong business performance once stewarded properly. When faced with establishing KPIs, Networks typically seek to focus on leading indicators. However, identifying the right KPI is not the only requirement from the Network. The Network must also assess the efficacy of processes and tools that may be in place to support the data collection for each KPI recommended for stewardship.

KPIs are not meant to be static; rather, they evolve over time based on the value created for the organization. For long-term success in KPI stewardship, Networks should identify and recommend six to nine KPIs for each PSM/Management System element. However, immediate focus should be on three to four high-value KPIs. As the organization focuses on these KPIs and begins harvesting the value from these KPIs, new and lower priority KPIs are introduced into the stewardship process. Tables 9.1 to 9.3 provide sample PSM KPIs that are useful to Networks for PSM elements. These KPIs are categorized according to the following:

- People (Table 9.1)
- Facilities and technology (Table 9.2)
- Processes and systems (Table 9.3)

TABLE 9.1

KPIs Related to PSM/Management System People

People Elements	Leading Indicators	Lagging Indicators
Leadership, management, and organizational commitment	• Number of field visits completed versus planned • Number of communication sessions completed versus planned	• Turnover rate
Security management and emergency preparedness	• Number of self-assessment audits completed versus planned • Number of security gaps identified • Number of emergency management drills completed versus scheduled	• Number of security breaches • Percent of gaps closed • Number of severe incidents requiring emergency response activation
Qualification, orientation, and training	• Percent of workforce competency assured • Percent of workforce trained • Percent of contractors oriented before arriving to site	• Number of incidents resulting from deficiencies in training and competency
Contractor/supplier management	• Number of joint contractor–owner leadership visits to the frontline versus planned • Number of contractor audits completed versus planned	• CRIF, CDIF • Number of recordable injuries • Number of contractor fatalities
Event management and learning	• Percent of high-priority incidents investigated • Percent of corrective actions implemented and closed • Average age of corrective actions	• Number of high priority incidents
Goals, targets, and planning	• Percent of work group with completed work plans • Number of performance evaluations completed versus planned	• Percent of work group without work plans

Notes: CRIF: Contractor recordable injury frequency; CDIF: Contractor disabling injury frequency.
Source: From Safety Erudite Inc., © 2012, https://www.safetyerudite.com/.

TABLE 9.2
KPIs Related to PSM/Management System Facilities and Technology

Facilities and Technology Elements	Leading Indicators	Lagging Indicators
Documentation management and process safety information	• Percent of standard operating procedures (SOPs) past due review dates	• Number of incidents related to absence of up-to-date process drawings
Process hazard analysis (PHA)	• Number of PHAs past cycle requirements	• Number of incidents that could have been avoided by a PHA
Physical asset system integrity and reliability	• Percent of overdue tests and inspection of PSM critical equipment • Number of bypassed interlocks • Number of bypassed alarms	• Number of incidents arising from mechanical integrity on PSM critical equipment
Operating procedure and safe work practices	• Number of spot audits done on procedure use and safe work practices	• Number of incidents related to absence of procedure and safe work practices
Legal requirements and commitment	• Number of regulatory incompliance	• Number of litigation cases

Source: From Safety Erudite Inc., © 2012, https://www.safetyerudite.com/.

TABLE 9.3
KPIs Related to PSM/Management System Processes and Systems

Processes and Systems Elements	Leading Indicators	Lagging Indicators
Management of engineered, nonengineered, and people change	• Total number of temporary changes • Total number of temporary changes past due dates	• Number of incidents arising from failure to complete a management of change (MOC)
Stakeholder management and communications	• Number of stakeholder engagement sessions conducted versus planned	• Number of protests arising from inadequate stakeholder engagement
Risk management	• Percent of work permits reviewed with incorrect hazards assessments	• Number of incidents arising from incorrect hazards assessments and risk ranking
Pre-startup safety review (PSSR)	• Number of startups completed without a documented PSSR	• Number of incidents arising from failure to complete a PSSR
Audits and assessments	• Number of self-assessments completed • Percent of audits not completed versus scheduled	• Number of incidents resulting from identified audit gaps
Management review	• Number of corrective actions issues related to KPI gaps	• Number of KPIs not being met • Number of KPIs not being followed up on

Source: From Safety Erudite Inc., © 2012, https://www.safetyerudite.com/.

Putting the Teeth into KPIs

KPI stewardship is not just another bit of work identified by the organization to be done but an important part of the organizational business process. The question therefore is: How does the organization get workers to take the stewardship of KPIs seriously? The answer is simple: KPIs must be linked to the rewards system of each employee. When a business performs well on its KPIs, employees are rewarded for their performance. On the other hand, when KPIs are not well stewarded and timely corrective actions are not initiated in response to deviations from planned performance, the business will ultimately perform poorly and employees will fail to maximize income and rewards from the organization.

In many large organizations, pay for performance as part of the rewards system is comprised of three components:

- A corporate or an organizational overall performance component.
- A business unit or major facility component (particularly if the business unit or facility operates as a component of the larger organizational entity).
- An individual or worker component, which is reflected in the work performance and behaviors of the individual employee as it relates to the KPI.

In many progressive organizations, improved performance targets are identified for each KPI the organization focuses on for stewardship. When KPIs are identified, ranges of performance targets are also allocated for each KPI that is linked to the overall performance of the business unit as well as the organization. Each KPI is weighted so that weighted scores are achieved for each KPI that is then used to determine the overall performance-related rewards for the employee at the business unit and corporate levels. These business unit and corporate KPI maximum, minimum, and target scoring ranges are communicated to all workers at the start of the new year. Table 9.4 provides a typical scoring process for a business unit and for the corporation.

At the end of each year, weighted performance scores are determined for the business unit and the corporation (the corporation may be made up of several business units with different KPIs; all of which contribute to the overall KPI stewardship and performance of the organization). These performances are then used to calculate the performance (bonus) pay received by each employee. A worked example is provided for your information in Table 9.5.

TABLE 9.4

Typical KPI Scoring Methodology for a Business Unit and Corporation

KPIs	Performance Range			Weight	Score	Weighted Score
	Maximum (200%)	Maximum (100%)	Minimum (80%)			
Percent of overdue tests and inspection of PSM critical equipment	0	1%–5%	>5%	1	200	200
Number of bypassed interlocks	1–5	5–10	>10	1	80	80
Percent of work permits reviewed with incorrect hazards assessments identified	0	<5%	>5%	2	100	200
Percent of startups completed without a documented PSSR	0	<5%	>5%	2	200	400
Percent of audits not completed versus scheduled	100	>95%	<95%	1	100	100
Percent of field visits completed versus planned	>100	90–100	<90	0.5	200	100
Percent of communication sessions completed versus planned	>100	90–100	<90	0.5	200	100
Percent of workforce trained	95–100	80-94	<80	1	100	100
Percent of workforce competency assured	95–100	80–94	<80	1	100	100
Weighted Score	Maximum: 2000	Target: 1000	Minimum: 800	10	1280	1380

TABLE 9.5

Pay for Performance Determination

KPIs	Contribution to Performance Pay (Determined by Corporate and Business Leaders) %	Weighted Score	Contributions to Worker Performance Pay (Actual KPI Performance/ Target KPI Performance)	Worker Performance Pay (Contributions from Corporate, Business Unit and Employee Performance) %
Corporate	30	1600	1.60	48.0
Business Unit	40	1380	1.38	55.2
Employee	30	Determined based on pre-established criteria with leader	1.00	30.0
Total	100			133.3

Worked Example of Performance Pay (Bonus) Linked to KPIs

Bearing in mind that business units and corporate KPI performance targets must be communicated to all workers early in the performance management year, the example provided in Table 9.5 is designed to provide an understanding of how employee performance pay or rewards (bonus) may be linked to KPI performance. Table 9.5 provides a simplified method for determining the employee bonus based on KPI performance.

As shown in Table 9.5, a worker will receive 33.3% of some number (generally base pay) as a bonus based on corporate, business unit, and personal contributions to KPI performance. In the example provided, a worker can potentially gain a corporate maximum of 60% and a business unit maximum of 80% that adds to his or her personal performance of 30%, which produces a bonus potential of 170%.

Target Setting

When setting performance targets related to KPIs, care must be taken to ensure attention is paid to the criteria identified in Table 9.6. Since performance on the KPIs is linked to pay for performance, there must be adequate transparency and defensibility to the process of setting targets and stewarding performance. The criteria listed in Table 9.6 help to achieve these goals.

TABLE 9.6

What Processes Are Used for Achieving Consensus around Targets

Criteria	Definition
What to target?	Which aspects of performance can most usefully be addressed via targets?
Measures to be used	What measures can most usefully be employed to define and assess progress? Clarity in both targets and measures are paramount to success.
Evidence to support performance, verifiable	What evidence is required to determine current status (where we are now) and future trends (where we want to be)? Is the evidence quantitative or does it depend on qualitative measures?
Process for consensus	By what process can consensus around targets be reached?
Who has accountability?	Who is accountable, have they accepted this accountability, and are they prepared to accept the challenges for achieving agreed targets?

Source: From the Royal Society for the Prevention of Accidents, 2002, Targets for change. Retrieved December 30, 2012, from http://www.gopop.org.uk/docs/targets.pdf.

Target setting, if it is to be useful, has to be based on good data, robust analysis, and a sound understanding of the processes through which improved risk management can be achieved. If targets appear to be plucked from thin air, not only will they lack transparency, meaning, and credibility, but they will not secure workforce and management buy-in and support. At a corporate or divisional level, where reductions in particular classes of accident and work-related ill health are being addressed through target setting, this needs to be based on a sound understanding of immediate and underlying causes and "preventability."

As with good budgeting, targets will not be robust unless they are based on a rigorous *ground up* approach in which, at each stage, the key stakeholders are subject to challenge on their estimates. In practice, however, output improvement targets (such as reductions in lost time incident frequency [LTIF]) may still have to be set on the basis of professional judgment. If this is the case, it is important that they are not simply a forward projection of some fraction of what has been achieved in the past, without assessing historically the relative contribution to such trends of factors, such as changes in hazard exposure of the workforce, as opposed to the impact of particular preventive interventions. Targets which merely assert unrealistic goals—but without indicating why they have been chosen or the means by which they were set or can be achieved—serve no useful purpose and need to be challenged. When establishing targets, organizations must therefore engage stakeholders and leverage Networks and Communities of Practice to gain credibility in terms of what gets measured.

Ensuring Relevance

In choosing targets that can help motivate and leverage change, Networks should seek to ensure that targets chosen are not only transparent but focus specifically on agreed priorities. In choosing priorities, they need to be able to balance the likelihood of achieving *quick wins* against making progress in solving intractable problems. For example, does the organization commit considerable resources to re-engineering a process to eliminate a remote risk with severe consequences or adopt a lower cost behavioral solution to deal with this problem and use the balance to address lower consequence but chronic problems that, so far, have proved difficult to solve?

Besides some sort of strategic appraisal or status review of the organization's PSM capability, target setting needs to be preceded by an analysis of its record in managing the principal risk exposures of the organization in the area of focus of the KPI. As with other forms of targets, the decisions about elements of the plan need to be determined through status review. They need to be evidence-based and progress toward them needs to be capable of being monitored.

Avoiding Distortions and Duplications

Apart from the danger already mentioned of simply setting targets for things that are easy to measure, there is the risk that inappropriate targets may cause key stakeholders to deliver against these measurable rather than real, underlying process requirements. Distortions can occur when unintended behavioral changes occur in response to a proposed KPI. For example, a KPI requiring buyers to use only prequalified contractors/suppliers in the bid and selection process may eliminate other competitive contractors/suppliers who may otherwise be unaware that a prequalification requirement exists for consideration. Such distortions may result in suboptimal value creation for the organization and may lead to buyer behaviors that fail to encourage and help contractors/suppliers to become prequalified. Thus, it is important that performance targets are only ever presented as a means to an end and must never be allowed to become an end in themselves. Furthermore, they need to be compatible (and not in conflict) with other performance targets set within the organization, otherwise achievement of improvement in one area will only be achieved at the expense of a deterioration in performance in others.

Duplicated KPIs not only place additional resource constraints on the organization, but they create confusion among stakeholders regarding who is doing what and who is ultimately accountable for the KPI. Therefore, when selecting, developing, and stewarding KPIs, there must be cross-Network engagement for overlapping PSM elements so that duplicated KPIs are avoided.

Consultation and Engagement

When setting targets (particularly for health, safety, and environment [HSE] purposes), employers have a statutory obligation to consult the workforce through trade union safety representatives or via safety committees. The important principle here is that the workforce, as keepers of the knowledge about working conditions and possibilities for change, need to be provided with opportunities to participate in discussions about current performance levels and possible future targets; and to compare notes with colleagues within and across their sectors as well as with others outside. Indeed, unless targets are developed with workforce involvement and buy-in, it is doubtful whether they will be achieved in practice.

Workers' perceptions with respect to health and safety at work are rarely taken into account when considering the development of such targets. In a focus group study to explore workers' perception to chemical risks, it was found that workers feel that their specific knowledge of their working conditions and their proposals for practical, cost-effective solutions to improve health and safety at the workplace are insufficiently taken into account. A participatory approach to target setting, however, may require new methods of working and new inputs, including, for example, closer partnership between workforce representatives, HSE professionals, and managers, and additional resources (Hambach et al., 2011). Beyond the workforce, organizations may also wish to consult more widely with outside stakeholders; for example, with business partners, with their insurers, with enforcing authorities, with unions nationally, with trade associations, or with benchmarking partners.

Similarly, where PSM KPIs are concerned, consultation and engagement is essential for the reasons identified earlier. Consultation of key stakeholders, such as SMEs and the Community of Practice, help the business in not only getting the KPI right but also provide for all criteria identified earlier in Table 9.5. Consultation and engagement where PSM KPIs are concerned help generate the right levels of ownership and enthusiasm regarding getting it right and generating strong business improvements for the organization.

Monitoring and Reviewing Progress

Although each target has a specific timescale for achievement, continuous monitoring and review is important in progressing toward stated targets. Remember the old adage "what gets measured gets done," therefore targets that are never subject to review and revision in the light of ongoing performance monitoring are also likely to be of limited value and are often forgotten. Feedback from interim review exercises should also include the possibility of changing or modifying targets.

If it turns out that the wrong targets have been selected, then they will need to be changed or adjusted, particularly if they are not achieving their objectives or are having a distorting effect on behavior. Besides helping to leverage improvement, much of the real value of target setting derives from the extent to which it helps those involved to develop their understanding of strategic management issues. From this perspective, understanding, in the light of experience, why a target was not reached or why it was in fact exceeded is actually what is of fundamental importance. Indeed, the challenge for all those involved in promoting health and safety in organizations is to ensure that the quest for continuous improvement in their subject area is addressed as part of the wider quest for business excellence.

Reporting against Targets

A healthy reporting culture is an essential part of a good Process Safety Management culture. It provides a valuable data source in the never-ending quest to reduce risk. Any performance report in a business context needs essentially to address the following:

- Where we were
- Where we said we wanted to be
- Where we are now
- Where we plan to be next
- How we plan to get there

Generally, performance reports will include reference to a range of key performance indicators. Whichever sort of corporate PSM KPI performance targets are chosen, reports on performance need to include commentary on whether these have been exceeded and the reasons in either case.

Challenges to KPIs

Selecting, developing, and stewarding KPIs is not without significant business challenges and constraints. Among common challenges encountered are the following:

- Participation issues
- Data management issues
- Observation issues
- Target setting constraints

Participation Issues

Participation drives ownership. In the absence of full stakeholder participation and engagement, ownership is weak among excluded stakeholders. In large corporations it is often quite difficult to engage and involve all stakeholders in the decision-making process. Care must be taken, therefore, to ensure that at the very minimum, communication to impact stakeholders is included in the process. The goal of participation in this instance is to avoid the treachery of *made by corporate and tossed over the fence to operations.*

Data Management Issues

The collection, storage, and processing of new data is always challenging in the first stages. There are numerous different KPIs for PSM processes, and each explicitly or implicitly requires some form of administration, data collection, and tabulation system. Where PSM and nontraditional KPIs are concerned, data collection, storage, and interpretation continue to be among the greatest challenges faced when stewarding KPIs. This is particularly so when data is not generally quantifiable and subject to the interpretations of an individual or group of individuals. Among the data management issues common to stewarding KPIs are the following:

- Point of collection of data—Where is the data collected?
- Who is responsible for collecting the data? The solution is to find a stakeholder closest to the point of collection of the data.
- What methods shall be used to analyze and interpret the data collected—Should be clearly defined and agreed upon by stakeholders before executing the KPI.
- Frequency of data collection—Is the data collection continuous, periodic, or point in time data?
- Quality of data collected—The level of difficulty and cost associated with collecting the data will also affect the quality of data collected.

Despite these challenges, once a KPI is established, developed, and accepted based on value creation in the organization, stewardship and ownership becomes easier and more sustainable among stakeholders as the organizational culture becomes impacted and the KPI approaches *value status* in the organization.

Observation Issues

Workplace observations are the key to success in identifying shortcomings in measuring KPIs put in place to achieve specific performance targets, as the data provide the means to give feedback and rectify any problematic safety issues. Where new KPIs are concerned, particularly as PSM is new in many regions of the world, strong leadership support is required to continually reinforce to workers and stakeholders that this is important to the organization and to your workplace. We need you to report on the measures communicated, and collectively we stand to benefit in the long run.

Target Setting Constraints

Implementing KPIs and performance targets to monitor and improve is often faced with problems arising from unduly taxing the business in the following ways:

- Time pressure constraints—Wanting to do things quickly without adopting an effective change management process to address impact on affected stakeholders.
- An underestimation of the magnitude of the undertaking and the failure to allocate sufficient resources to do a good job.
- Poorly articulated KPIs—Doing the work upfront to ensure the right KPI is executed with the right type of data collected at the right frequency has a greater chance of success than KPIs where we fail to do this right. Getting it right the first time helps to institutionalize the KPI among stakeholders to generate long-term ownership and success.

Conclusion

The development of KPIs to enhance PSM is a worthwhile venture. The setting of performance targets and tracking them with KPIs is the best way for a company to comply with and exceed industry standards. It also has the added benefit of, if used correctly, boosting a company's public image and increasing profit margin by reducing the downtime associated with health- and safety-related issues that could have been avoided. Benchmarking enables the company to leverage itself among the key competitors and work toward improving its own processes. Recent study has shown that the enforcement of safety regulations on major chemical accident control have not been satisfactory, mainly due to lack of awareness and weak management in the industry. The proposed system and framework for PSM KPIs and performance targets can support the efforts to improve the safety management in any industry. Last, and by no means least, Networks and Communities of Practice provide a credible framework of structure, expertise, and engagement for developing the right KPIs for sustainable business performance.

References

Broadribb, M. P., Boyle, B., and Tanzi, S. J. (2010). Cheddar or Swiss? How strong are your barriers? (One company's experience with process safety metrics). *Loss Prevention Bulletin,* 212: 29–40. Retrieved December 30, 2012, from EBSCOhost database.

Brooks, W. K., and Coleman, G. D. (2003). Evaluating key performance indicators used to drive contractor behavior at AEDC. *Engineering Management Journal,* 15(4): 29–39. Retrieved November 12, 2012, from EBSCOhost database.

Fishwick, T. (2010). Key performance indicators. Signposts to loss prevention. *Loss Prevention Bulletin,* 212: 4–5. Retrieved December 30, 2012, from EBSCOhost database.

Hambach, R., Mairiaux, P., Francois, G., Braeckman, L., Balsat, A., Van Hal, G., Vandoorne, C., Van Royen, P., and van Sprundel, M. (2011). Workers' perception of chemical risks: A focus group study. *Risk Analysis,* 31(2): 335–342.

Plastics and Chemicals Industries Association (PACIA). (2010). Guidance document: Process safety—Developing key performance indicators. Retrieved December 30, 2012, from EBSCOhost database.

Royal Society for the Prevention of Accidents. (2002). Targets for change. Retrieved December 30, 2012, from http://www.gopop.org.uk/docs/targets.pdf.

Safety Erudite Inc. (2012). Retrieved from https://www.safetyerudite.com/.

10

Challenges Faced by Organizations in Managing Networks

For most of the organizations establishing Networks in the workplace, this is a fairly new concept. The ways Networks are expected to function and operate are now being developed, and organizations are wrestling with issues and challenges around the following:

1. Size of the Network—What is the optimal size of the Network?
2. Representation—Should representation be limited to business units or extended to business functions and facilities?
3. The right balance between face-to-face meetings and virtual meetings.
4. Quantifying the value created by the Network.
5. Understanding the right balance between Network responsibilities and full-time duties.
6. The cost associated with travel, accommodation, and international commute for globally distributed Network members.
7. The numbers of Networks to be established.
8. Managing the interrelationships between overlapping Networks.

When organizations attempt to rollout Networks across their business, many understand *what* is required from these Networks but lack the understanding of *how* to make them work effectively. This chapter examines challenges faced by organizations in setting up Networks and will attempt to provide some guidance in addressing these challenges.

Size of the Network: What Is the Optimal Size of a Network?

A crucial requirement for ensuring the success of a Network is the right sizing of the Network. By right sizing, we mean the numbers of members in the Core Team, support subject matter experts (SMEs), and within the

Community of Practice. Several factors will influence the size of the Network, among which are the following:

- Size and scale of operation of the organization.
- The level of maturity of the organization with respect to the Network being formed.
- The numbers of stakeholders involved in the area of focus of the Network.
- The amount of work that must be undertaken to close gaps between current and desired states.

Size and Scale of Operation of the Organization

The larger and more global the organization, the larger the size of the Network. In global organizations, it is more important to have larger SME pools and Communities of Practice that serve as conduits for information and knowledge transfers and for getting best practices and new work processes executed at the frontline. The Core Team, on the other hand, should generally be limited to the six to eight members that include the major stakeholder interests. The smaller the Core Team, the easier it is to manage while drawing upon the expertise that resides within the SME pool.

Level of Maturity of the Organization with Respect to the Network Being Formed

When organizations are relatively mature in a particular operating process, Process Safety Management (PSM) element, or management system element, the potential gaps to be closed may be few and small. Networks established under such circumstances may be relatively small and Network focus will generally be primarily on continuous improvements. Many large and mature organizations that have introduced some form of a management system will generally have some systems, processes, and tools in place to meet the requirements of the PSM or Management System element requirements. In such organizations, the gaps between the current and desired states may be significantly smaller than that in less mature organizations.

Networks formed under these circumstances function more as a body aimed at continuous improvements. The goal of such Networks tend to focus on looking for better and improved ways of doing work within the organization as opposed to developing tools, processes, and systems, which is characteristic of less mature organizations. In view of this, the maturity of the organization will influence the size of the Network across all levels of the Core Team, SME pool, and Community of Practice.

TABLE 10.1

Stakeholder Groups Associated with Two Different Networks

	Networks	
Stakeholder Group	**Contractor Safety Management**	**Management of Change**
Business Unit	√	√
Engineering		√
HSSE	√	√
Legal	√	
SCM	√	
Communications	√	√
Training	√	√
Change Management	√	√
Risk Management		√
Total	7	7

Number of Stakeholders Involved in the Area of Focus of the Network

The numbers of stakeholders and stakeholder groups will naturally influence the size of the Network. Table 10.1 compares the different stakeholder groups likely to be involved in two different Networks.

Amount of Work That Must Be Undertaken to Close Gaps between Current and Desired States

A gap analysis that identifies the differences between the organization's current and desired states will determine the amount of work to be accomplished before the organization can be placed onto a sustainable platform. Large gaps and the increasing numbers of gaps suggest a larger amount of work to be performed by the Network. In such situations, the size of the Network may be greater such that work can be allocated to various Core Team members and working subgroups. The size of the Community of Practice may also vary based on the amount of frontline work required to execute required changes at the frontline.

Representation: Should Representation Be Limited to Business Units or Extended to Business Functions and Facilities?

Having the right representation will help with ownership of the process. Various functional organizations and support services are necessary requirements for the effective functioning of Networks. Furthermore, when certain facilities possess superior work practices, the change impact is likely to be high from gap closure strategies suggested by the Network, and such stakeholder representation is invaluable.

Undoubtedly, therefore, Networks must be extended to include business functional groups as well as facilities. A good example of the need for stakeholder representation is reflected in a Contractor Safety Network. Ensuring stakeholder representations in the Core Team is essential for full buy-in and support. When standardization of the corporate contractor prequalification process is identified as a Network deliverable, the following key stakeholder groups should be considered for inclusion in the Core Team:

- Business unit—Each business unit of the organization should be allowed representation on the Core Team since there is generally huge vested interest in ensuring both their specific needs are met and unique challenges and concerns are addressed.
- Supply chain management (SCM)—SCM will generally own and administer the process. There will be weak ownership and support should SCM not be represented within the Core Team.
- Legal—Legal representation is required to ensure the organization does not assume unintended liabilities from recommendations and actions suggested by the Core Team.
- Health, safety, security, and environment (HSSE)—Since Contractor Safety is the basis for the Network, HSSE representation is paramount to the success of the Network. Failure to include HSSE representation in the Core Team will lead to a dysfunctional Network.

Therefore, when setting up Networks, getting the Network size right, with the right representation, helps in both the decision-making process as well as in securing ownership and buy-in for suggested solutions and recommendations for gap closure strategies.

Right Balance between Face-to-Face Meetings and Virtual Meetings

Much of the work performed by Networks is generally undertaken by virtual teams given the global and geographical spread of organizations today. Striking the right balance between virtual work and face-to-face meetings is an important criterion for the success of Networks. Although virtual technologies serve to help steward Network progress on an ongoing basis, the inclusion of face-to-face meetings helps to address complex issues that may reside within the deliverables of the Network.

This is critically important when a diverse Core Team is established that comprises varying cultures. The ability to recognize body language during face-to-face meetings is extremely important in determining whether stakeholder's interests are satisfied. Furthermore, when the Core Team meets during face-to-face encounters for 2 to 3 days, a significant amount of work is accomplished.

The concern associated with face-to-face meetings of the Networks is balancing cost with value created from face-to-face meetings. Bearing in mind that members of the Core Team may be geographically dispersed, with several functioning Networks, cost of travel and accommodation can quickly climb to astronomical levels, and senior leaders would necessarily want to understand value creation versus cost of doing so, which brings us to the next challenge of how do we quantify the value generated by the Network.

Quantifying the Value Created by the Network

Quantifying the value generated from Network activities is an extremely difficult exercise. Value created by Networks can be expressed both in quantitative and qualitative terms. Many organizations can become fixated and bogged down with debating decimal point accuracy of the value created by Networks—conditions that should be avoided and mitigated at the senior leadership levels.

Tangible or Quantitative Assessments

Quantitative expression of value created easily justifies the existence of a Network. Value created is easily expressed in a dollar amount. For example, a rotating equipment Network may recommend the standardized use of a specific type of pump that increases the intervals between preventative maintenance by 50% and reduces pump seal failures from four times per year to two times per year. Value created from reduced maintenance cost and from extended process runs can be easily quantified and extrapolated over multiple units where applicable to demonstrate the value created from such Networks.

Intangible or Qualitative Assessments

Qualitative assessments and quantification of value created from Networks is much more difficult to justify. For example, the introduction of a new contractor prequalification process by the Contractor Safety Network may result in a reduction of the contractor recordable injury frequency (CRIF) from 1.2

to 0.8. From a dollars and cents perspective, however, it is very difficult to quantify how much was saved by the organization from the introduction of the prequalification process. Furthermore, when several other Networks are contributing to this improvement, it may become an even more difficult task to attribute the improvements derived from the prequalification process from that associated with, say, improvements in training and competency assurance delivered by the Training and Competency Network.

The most important principles that should be followed for identifying value contributed by Networks include:

1. The operating business units own the dollar value improvements (their bottom line)
2. Value creation is not an accounting exercise to keep score
3. Avoid calculating value created to the decimal points
4. Avoid allocating percentages value creation to contributing functional and supporting departments
5. Celebrate success and provide recognition for the Networks as value contributions are identified

Many organizations can get bogged down with debating decimal point accuracy in items 1 to 4. It is more important for the organization to ensure visibility of the gross value, both tangible and intangible, being created by the Network activity. The visibility of Network-contributed value is a critical success factor for Network sustainability.

Understanding the Right Balance between Network Responsibilities and Full-Time Duties

When Networks are established they are generally set up with Core Team members performing Network-related duties along with the principal duties of the members' substantive roles in the organization. Over time, if not managed well, and in situations where the extent of gap closure requirements is extensive, the demands on Network members can impact their work performance in other areas.

Therefore, care must be taken to ensure the workloads imposed on workers who function as Network members are reasonable and achievable. More important, members are allowed the right levels of work–life balance while functioning as a Network member. This is particularly important for Core Team members where burnout is likely to occur with extensive workloads.

TABLE 10.2

Network Membership Time Allocated to the Network Work Deliverables

Network Role	Percent of Work Plan Allocated
Network leader	25%–35%
Core Team members	15%–25%
SMEs	10%–15%
Community of Practice member	5%–10%

Table 10.2 provides a typical estimate of the Network member's annual work plan allocated for Network-related work.

Although the commitment requirements for most Networks may be met via the aforementioned allocations, some Networks may require Network leaders that are fully allocated 100% of the time to leading the particular Network. Such Networks may focus on high value opportunities for the organization. When full-time Network leads are assigned, there may be a fixed duration assigned to such roles, such as 2 to 3 years, to accomplish major deliverables. Once major deliverables are achieved, Network leadership can once again revert to part of a role requirement as per Table 10.2.

Cost Associated with Travel, Accommodation, and International Commute for Globally Distributed Network Members

Much like any business process or project management activity, cost management is very important to the success of a Network. The Network leader should be responsible for an allocated budget and should be held accountable for managing the budget. Budgetary constraints will help the Network leader develop the right balance for various Network activities, such as face-to-face meeting frequencies and the amount of travel required to deliver on assigned Network assignments.

Number of Networks to Be Established

Leadership must seek to establish the right balance between the number of Networks required by the organization to deliver on its strategic vision. We recommend that organizations evaluate the business needs for Networks based on the following:

1. Risk exposures presented to the organization
2. Value creation opportunities to be derived from setting up a particular Network
3. Overall business impact for establishing the Network

Organizations with limited experience in managing Networks should set up a few initial Networks to gain experience and learn from these experiences.

Ideally, organizations should initiate four to six Networks as a prekick-off exercise, applying the management process required to promote success of these teams. With learning and knowledge acquired from these initial Networks, a Network conference should be considered for kicking off all other essential and required Networks. Over time and as the Network process matures in the organization, a screening process should be instituted to ensure only required Networks are activated.

Managing the Interrelationships between Overlapping Networks

Where PSM is concerned, there are significant areas of overlap among many PSM elements. By extension, therefore, when Networks are established there should be similar overlaps among these Networks to promote value maximization and remove duplication of efforts. Networks' interrelationships present a unique problem for senior leadership when managing Networks. A strong hand of authority is required to help determine in scope and out of scope items for a particular Network. For example, the relationship between a Management of Change (MOC) Network and a Training and Competency Assurance Network can be quite challenging regarding the execution of training associated with change.

Issues such as the following are among the many that may arise when Network responsibilities are not clearly defined and understood:

- Which Network prepares and owns the training materials?
- Who executes and conducts required training?
- Where does cost of training reside?
- Medium for training—Who determines this? For example, who determines whether computer-based training is preferred over instructor-led training?

In view of potential conflicts arising from Network activities, organizations may be tasked with ensuring a strong Network governing body is in place to supervise and monitor the work of Networks.

11

Network Coordination: The Leadership Challenge

Senior leadership support and engagement is perhaps the most important requirement for Network success. Therefore, Networks must be supported with the right organizational structure such that the right balance between autonomy and control is provided to deliver on the charter deliverables. Moreover, an organizational structure helps in the removal of obstacles to progress and provides the required authority for enabling Networks. This chapter presents a leadership structure and model that is capable of enabling Networks to succeed.

Figure 11.1 provides a tried and tested organizational structure for coordinating the work of Networks and for Network success. Key linkages and reporting requirements for success include the following:

- Reporting relationship and links through the role of senior vice president of health, safety, security, and environment (HSSE).
- Business unit leadership involvement and support with direct reporting relationship with Network.
- Functional leadership involvement and support with direct reporting relationship with Network.
- A Network Steering Team for supporting and enabling all Networks.
- The Core Team of the Network.

Beyond the linkages and reporting relationships is the critical need for the following:

- Coordinated communication among Networks.
- Network reporting and stewardship.

Role of the Executive Vice President of Health, Safety, Security, and Environment

It makes good sense to have Process Safety Management (PSM) Networks and Networks in general report to the executive vice president (EVP) of

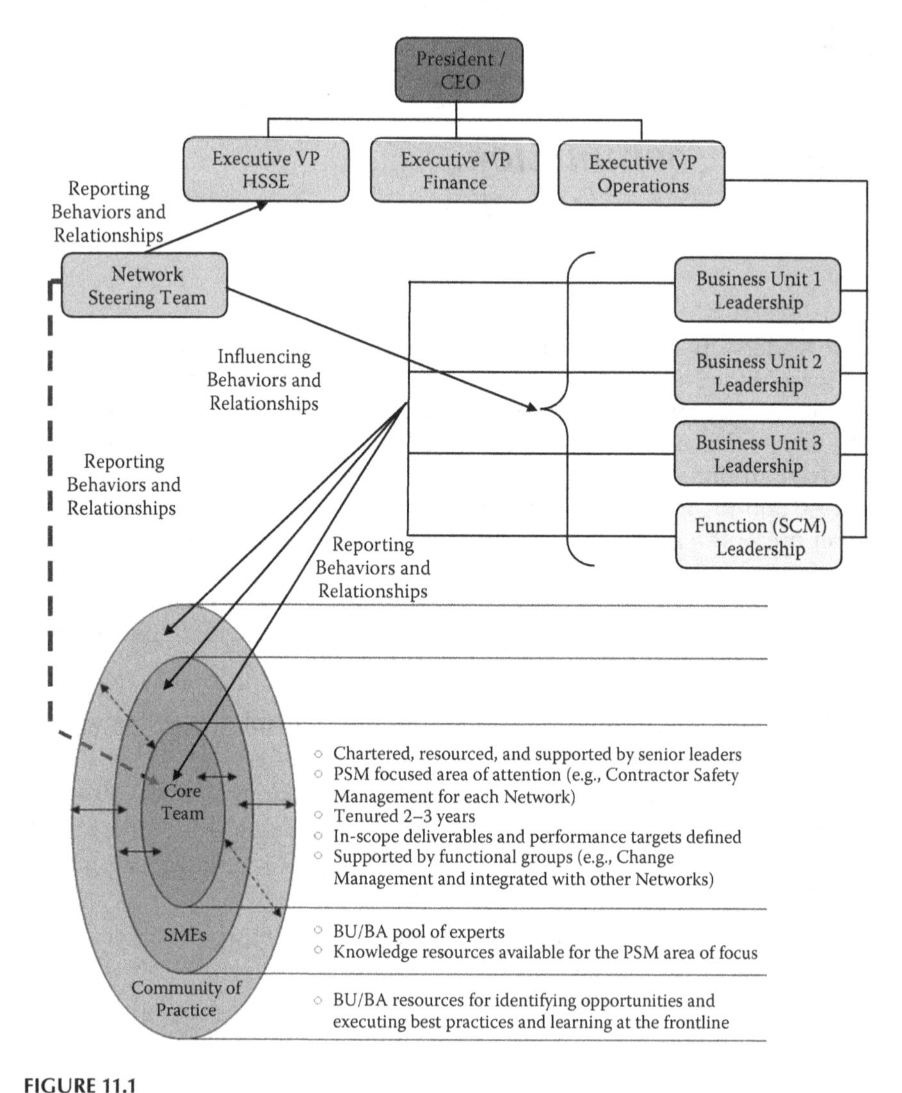

FIGURE 11.1
Organizational structure for Network coordination. (From https://www.safetyerudite.com/. © 2012 by Safety Erudite Inc. With permission.)

HSSE because of the health, safety, and environmental impact and focus of PSM Networks. An executive leadership requirement is essential to ensure that Networks receive the levels of prominence required by the organization. The EVP helps in setting the strategic direction of Networks, and securing required and essential resources to achieve the annual and strategic PSM goals of the organization.

Executive level support helps in securing the right levels of organizational support for Networks such that resources can be properly allocated and managed in achieving Network deliverables. From an optics viewpoint, a high-level reporting structure provides Networks a clear line of sight with the highest authority of the organization—the CEO.

Role of the Network Steering Team

The Network Steering Team functions as a stewarding body for Networks. The Network Steering Team is comprised of three to five senior leaders who are authorized by the executives of the organization to manage and steward the performance of Networks. The Steering Team executes the following duties:

- Works with Networks to ensure approved priorities are aligned with corporate vision and strategies.
- Reviews and approves Network charters.
- Secures and provides the necessary resources for the Networks to execute the deliverables.
- Reviews and approves the priorities and significant impact recommendations from the Networks.
- Supports performance of Networks, taking corrective action as necessary.
- Authorizes and approves new Networks.
- Communicates Network performance and significant achievements to the executive leadership of the organization.
- Communicates Network success and priorities within their functions and departments to ensure visibility of Network activity.
- Reviews Network charters and deliverables to eliminate duplicated work and remove confusion from overlapping priorities.
- Removes obstacles to the smooth functioning of Networks by interfacing with business unit leaders as required.

The Network Steering Team maintains a meeting frequency that provides sufficient direction to the Networks. These meetings serve to set priorities, review recommendations from the Networks, review incident learnings and best practices, and review the Network stewardship reports. Suggestions for new scope activities will also be brought to this team for discussion and potential resourcing. Steering Team members represent the primary impact areas of the process safety improvement work. These steering committee members also need to have the resource approval authority necessary to enable the Networks to execute their responsibilities.

Business Unit Leadership

Generally, Network members, subject matter experts (SMEs), and members of the Communities of Practice reside within the business units. Business unit leaders must support the release of Network and Community of Practice members so that they can contribute in a meaningful way to the Network deliverables. Typically, core members may commit 15% to 25% of their time to Network deliverables. Business unit leadership should therefore provide the framework for enabling Network members such that they are motivated to continue to remain focused on Network deliverables.

Business unit leaders work collaboratively with the Network Steering Team to address and remove obstacles that may impact the delivery of Network deliverables. In so doing, among other support provided, business unit leaders should do the following:

- Release Network contributors as required.
- Identify and support Network SME contributors' engagement and involvement with the Core Team of the Network.
- Relentlessly identify and share operating practices and procedures with the Network so that comparative assessments can be made among similar operating practices and procedures from other business units to identify potential best practices.
- Support the execution of standardized and improved business practices and procedures, drive for standardization wherever possible.
- Manage required change so as to minimize impact to all stakeholders, institute change management process to ensure all stakeholder risks and challenges are properly addressed and mitigated.

Functional Unit Leadership

In many instances the work of Networks cannot be completed without the support and involvement of functional units such as supply chain management (SCM) or human resources (HR) management. Similar to business unit leadership, leaders of the functional units are required to enable Networks by providing competent and motivated subject matter experts to support the delivery of assigned priorities.

Corporate or Central Control of Networks

Core Teams of Networks as well as the Network Steering Team should reside within the corporate organization. Among the most common reasons for central control of Networks are the following:

- Centrally controlled Networks are closer to the core of decision making for the entire organization, and therefore critical decisions are more easily fast tracked as required.
- Corporate or central control removes the risk of business unit influence on Network deliverables, for example, selecting the work practice or procedure that resides within the business unit as opposed to another more preferred option from another business unit (or even external to the organization) to avoid the change management requirements.

Figure 11.2 provides an overview of the central control required for Networks.

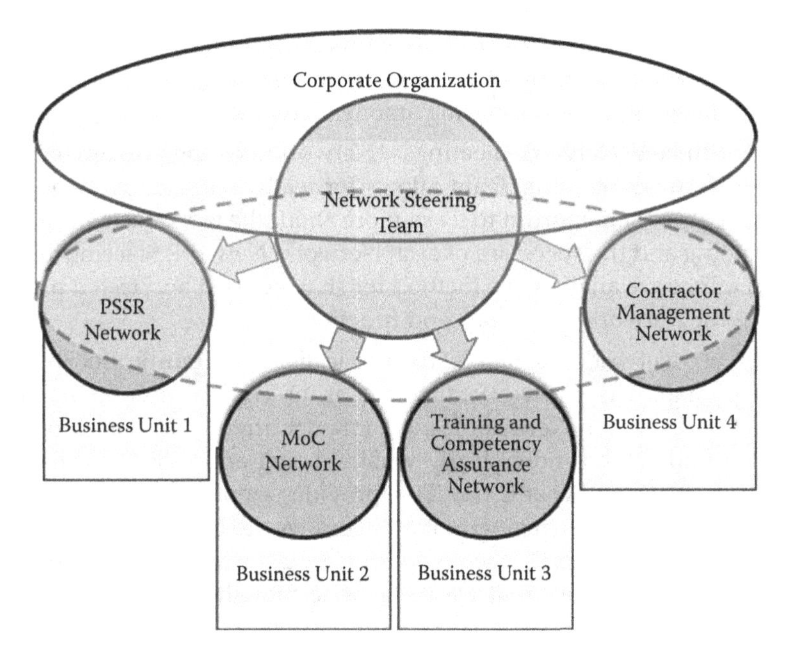

FIGURE 11.2
Central or corporate control required for coordinating Networks. (From https://www.safetyerudite.com/. © 2012 by Safety Erudite Inc. With permission.)

Network Communication

Because many elements of PSM overlap each other in the areas of people, processes and systems, and facilities and technology, long-term success of all PSM Networks require excellent internal Network communication as well as communication between and among Networks. Each Network must therefore be aware of the work being performed by each other, and the Core Teams of each Network must have access to each other's work.

Various methods are adopted to foster Network–Network communication. Communication among Networks must be organized such that maximum value and collaboration can be derived from the communications methods adopted. Among the more accepted methods for communication and collaboration are the following:

- Use of collaboration sites—Since all Networks may not all be in the same place at the same time, among the easiest method for Network–Network communication is the use of collaboration sites within the organization. Collaboration sites allow Core Team members from geographically distributed regions and time zones to work off the same platform on an ongoing basis. Collaboration sites also help Networks in cost management since multiple Network members can work off the same site at the same time. Collaboration sites for each Network with open/read access to all sites by each member Network facilitates information sharing among Networks.
- Coordinated Network meetings—Network meetings organized by the Network Steering Team allows for leaders of each Network to meet at the same forum to learn more about the work each Network is doing and the successes of each Network. Network Steering Team meetings are an effective method for sharing tried and tested methods for Network operations and functioning.
- Direct Network leader–Network leader communication—This is perhaps the most effective means of communication among Networks. When Network leaders interact directly with each other, they understand more clearly what is being addressed within the Network versus what is not. This provides ample opportunities for discussions and ensuring that all activities are being addressed by the various Networks, thereby eliminating the potential for missed work by a Network that was otherwise thought to be handled by another Network.

Figure 11.3 provides a simplified model of communication among various stakeholder groups for Network functioning. As shown in the model:

- Key messages with major change impacts are communicated to senior leadership and business unit leaders via the Network Steering Team.
- Continuous two-way communication between the Network and business unit stakeholders is required.
- Networks communicate with each other on a continuous basis.
- Networks communicate with the Steering Team on a continuous basis.

When major changes impacting all stakeholders are concerned, such communications are best executed via a face-to-face forum or conference so that all stakeholders hear the same message at the same time and interpret the message consistently.

Communication among Networks and all stakeholders is of tremendous importance to stakeholders since progress updates on deliverables that are important to the business as well as any upcoming changes are shared with

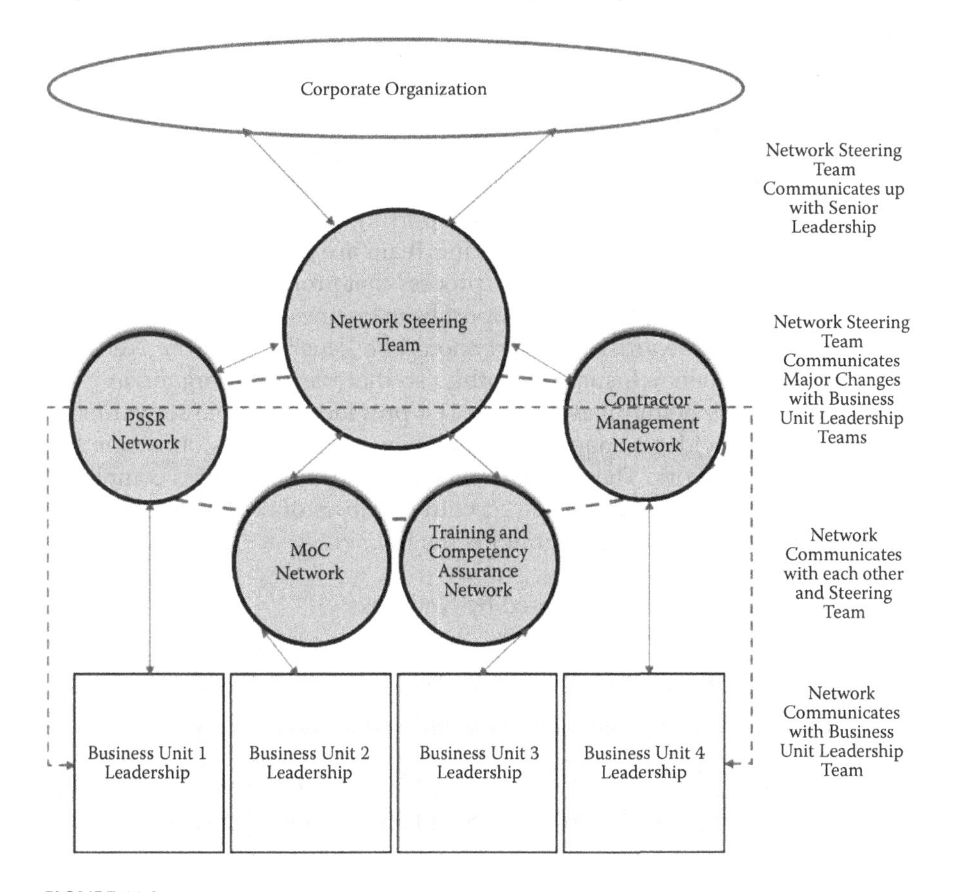

FIGURE 11.3
A communications model for coordinating Networks. (From https://www.safetyerudite.com/. © 2012 by Safety Erudite Inc. With permission.)

the business. In so doing, the business is more prepared for the next phase of the process for execution and implementation of proposed improvements.

Network Stewardship and Performance Management

Network stewardship in this instance can be explained as the required self-assessment and performance measurement part of the Network governance process. Each Network will have its own metrics and report them monthly to the Steering Team, using a standardized format or template. The details of what is measured will be different for each Network. Networks will be measured by their results; however, they will also be measured via the use of in-process or leading indicator performance metrics.

Networks, like any other organizational group, must be stewarded and supported to ensure the full potential of the work group. The need is even greater for Networks, which largely operate autonomously with guidance and control from the Steering Team. Since Core Teams of Networks are generally distributed within the business units and corporate environments coming together at meetings to discuss deliverables, the need for stewardship is vital to success. More important, the reporting relationships between the Network leader and the Network Steering Team are generally nonauthoritative, thereby requiring a stewardship process that promotes enablement.

In order for Networks to retain support from business unit leaders and the Steering Team, a stewardship report should be generated by the Network at an agreed frequency (usually monthly) so that the entire organization is aware of progress being made. Figure 11.4 provides a sample stewardship template that provides a snapshot of the Network's progress on a monthly basis for each Network. The template one-page type document is completed by the Network leader, then reviewed with members of the Core Team.

This stewardship template highlights the following:

- Major changes being proposed by Networks
- New practices or standards that are to be introduced organization-wide
- Work in progress
- Successes and value being contributed by the Network

The stewardship template also includes data that can be used by the Steering Team to assess Network health and effectiveness. As such, information such as the following is also included in the stewardship report:

- Number of meetings being held by the Network during the reporting period

Network–MoC Lead:	**Date:**
Overall Status	• On target. • Small manageable delays. • Needs help from Steering Team.
Sub-Initiative Updates(s)	• Clearly identify the status of Network activities and priorities generated from the opportunity matrix. • Provide updates on all sub-initiatives being worked by the Network.
Next Period Outlook	• Provide a listing of all major upcoming activities projected by the Network.
Support Requirements	
Steering Team	• Identify support required from the Steering Team by the Network.
Issues and/or Challenges	• Provide a list of challenges faced by the Network that may require senior leadership or business unit leadership support for addressing them.
Network Successes	
• List all achievements and successes delivered by the Network.	
Network Participation	
Meeting attendance	
% of Network absenteeism	
# of meetings cancelled	
# of Network vacancies	

FIGURE 11.4
Sample Network stewardship template. (From https://www.safetyerudite.com/. © 2012 by Safety Erudite Inc. With permission.)

- Percent attendance by Core Team members
- Number of meetings canceled due to poor attendance and other controllable factors
- Number of Network vacancies and turnover rates within Networks

Trends can be assessed, and appropriate follow-up or course correction can be instituted accordingly.

Value creation contributed by the Networks is also tracked in the stewardship report to reinforce and celebrate success. Networks contribute tangible and intangible value in many ways, among which are the following:

- Reducing risk, thereby avoiding losses to the organization

- Acceleration of the use of best practices across the entire organization
- Introduction of standardized work practices and procedures across the entire organization
- Building competence within the organization by capturing and transferring knowledge and learnings across the entire organization
- Accelerating improvement as demonstrated through increasing revenues or reducing costs

Highlighting value creation of Networks is an important aspect of Network stewardship since formal Networks as centers of excellence in organizations is a relatively new concept that is yet to gain widespread acceptance across organizations.

Conclusion

DuPont has demonstrated, in no uncertain terms, that when executed and supported properly across organizations, Networks can bring about improvements to business. Through the lenses of a dynamic leader in electrical safety and technology at DuPont who was dissatisfied with the numbers of electrical technician workplace fatalities in the organization in the pre-1992 period, an opportunity for an Electrical Safety Network within DuPont arose. Since 1992, due to the efforts of this Network, DuPont has sustained a zero fatality record from electrical incidents. Prior to the introduction of the Electrical Safety Network in DuPont, the organization experienced a fatality rate of one fatality every 3 years.

Networks function with a fair amount of autonomy and are often led by Core Team members who demonstrate a type of passion for improvements in the particular area of focus of the Network. To harness this passion and channel it into generating positive business outcomes is an often-challenging task. However, Networks are critically important to businesses in today's business environment that is characterized by scarce and dwindling resources and must therefore be introduced and managed to generate continuous improvements to businesses.

The Network Steering Team has a tremendous leadership challenge resulting from nonauthoritative leadership control across multiple Networks generally of different maturity levels. More important, since Network deliverables are generally in addition to the normal day-to-day routines of the Network members, exceptional transformational leadership skills and behaviors are required by the Steering Team members to inspire the hearts and minds of the Core Team members. Demonstrated commitment, leadership behaviors, and values supporting the vision and goals of each Network are essential.

The challenge faced by business unit leaders continues to be the same of doing more with less. Indeed, as Networks become more productive and start generating value to the entire organization in the form of standardized work practices, which generate synergies and value and the introduction of best practices, which generate greater efficiencies and effectiveness—the concept of doing more with less goes away. However, until such time that value is created from Networks the challenge of doing more from less remains a burden to business unit leaders.

Finally, communication among Networks and other key stakeholders continues to remain a leadership challenge for the Steering Team. While organizations may introduce collaboration sites and technology for improving communication among globally dispersed Core Team members, Generation X members continue to avoid such tools and processes. However, experience and knowledge in many PSM Network focus areas continue to reside in the experience and knowledge of older workers who have gathered experience and knowledge over extended periods. Often, Core Team members revert to face-to-face meetings, which pose intense cost pressures on the organization.

12

A Network at Work

This chapter provides an example of a working Network that utilizes the processes and tools discussed throughout this book. The example demonstrates how to use the tools and templates required for the Network to work effectively. Recapping the key requirements for setting up a Network, the following topics will be discussed in this chapter:

- Determining the need for Network support
- Example of a Network charter
- Example of prioritization of work plan items by using the opportunity matrix
- Example of a work plan
- Communication plan and messaging
- Stewardship and communicating up the organizational chain
- Execution of proposed improvements

Determining the Need for Network Support

When the strategic goal is determined by senior leaders of the organization to improve reliability and reduce unplanned outages, management of change (MOC) becomes an important part of the process along with other Process Safety Management (PSM) elements such as mechanical integrity (MI), pre-startup safety review (PSSR), and quality assurance (QA). An internal evaluation of the organization's MOC historical performance will guide the business and the Steering Team in determining whether a MOC Network is required. At the business unit level, alignment with the strategic goal of the organization resides in the following performance criteria:

- A reduction in risk of catastrophic incidents (i.e., fire or explosion).
- An improvement in reliability of plants or part of a plant in various locations of an organization.
- Standardization across the entire organization and across multiple plants in various geographic locations.

- Teamwork and subject matter experts required to develop standards, guidelines, tools and templates, training, communication plans, and audit protocols to bring about required improvements in performance and for achieving the strategic goals of the organization.

Compelling support from business units highlighting performance gaps may lead to MOC, PSSR, and MI Networks. Once the need for a MOC Network is approved by the Steering Team, a Network is established leveraging capabilities from the business units of the organization. Recommended membership composition should include the following:

- A corporate MOC specialist, Network leader
- A change management and communications specialist to help execute process and technological changes across the organization
- Access to training and competency assurance resources
- Management of change experts from various business units
- A health, safety, security, and environment (HSSE) professional

Developing the Management of Change Network Charter

As discussed earlier, a charter helps to keep the Network focused on the deliverables and goals of the organization. Once the Core Team has been identified and given the strategic goals of the organization regarding MOC, the Network members must work together to clearly define the deliverables and what is in scope and out of scope for the Network. By working together with the Steering Team and business leader, the Network is better able to understand the focus of the organization and to develop strategies for achieving the strategic goals of the organization regarding MOC.

All Network team members must be involved in finalizing the team charter. The charter must clearly define what the deliverables are and what is in scope or out of scope for the team, and all members of the Core Team must agree to this. Once the Core Team approves the charter, the Steering Team provides oversight and review and may consider adding or removing items from the charter based on potential overlap with other Networks' efforts. Table 12.1 provides a working example of a MOC charter.

TABLE 12.1

Sample of Completed MOC Charter

Purpose: Network the (Company Name) wide community that deals with management of change (MOC, excluding MOC–People) to drive operational excellence. Do so while leveraging the knowledge and expertise that resides within the organization and throughout the industry.
Timeframe: Steering Team to review Network effectiveness monthly. *Commencing:* (Insert Kick-Off Date Here)

Objectives	Measurable Goals	Key Activities	Tracking Measures (How)	Critical Success Factors
• Support/drive operational excellence in management of change (MOC) • Compare, contrast, reconcile best internal operating practices between business units and business areas across the Network • Identify areas of opportunity for continuous improvement of the MOC standards • Support implementation of the standards across the Network • Provide a forum to facilitate these objectives efficiently and effectively • Provide guidance to the Network on the interpretation of standards • Build competencies	• Effective usage of MOC processes across the organization • Increased conformance to MOC standards across the organization • Maintain and continuously improve standards by regularly engaging the Network • Timely response to questions/requests from the business	• Monitor the current state of MOC across the organization • Define the MOC metrics for improving MOC performance • Drive the improvement of the standard, definitions, and audit protocol documents • Seek out best practices from across organization and industry for organizational sharing • Drive the type and quality of MOC training required • Participate in PSM audits relating to MOC as needed • Interface and collaborate with other Networks	• Audit and self-assessment of performance • Improvements in performance on corporate and local MOC KPIs and metrics • Reduction in turnaround time for support/responses to queries from business and Communities of Practice • Demonstrated behaviors endorsing the use and application of the MOC standard and its intent	• Results oriented • Team composition • Business unit approval and support of members' commitment • Effective steering/facilitation • Excellent communication and visibility • Clear performance measures • Membership and performance of the members in their personal goals and scorecards • Best in class infrastructure to support efforts • Collaborating with other Networks

(continued)

TABLE 12.1 (CONTINUED)

Sample of Completed MOC Charter

Objectives	Measurable Goals	Key Activities	Tracking Measures (How)	Critical Success Factors
required to perform MOC in the Communities of Practice • Reduce number of severity of process safety incidents related to MOC.				
In Scope	**Out of Scope**	**Deliverables**	**Members**	**Meetings**
• Assist in revision and continuous improvement of MOC standards • Development and enhancement of training material (generic, not site specific) • Engagement with the Network covering all business units and areas • Engagement with other Networks as appropriate • Key stakeholders for MOC (IT) tool development/ sustainment • Inclusion of Management System requirements for MOC	• MOC people • Site Implementation of MOC or other site-specific initiatives • Site selection of MOC personnel • MOC IT tool implementation and sustainment	• An up-to-date MOC standard, audit protocols, PSM definitions • Up-to-date training materials • Monthly stewardship summaries of activity and results	• Sponsor • Team lead • Team members	• Notionally 12 meetings per year • 1 face-to-face annually • 11 virtual meetings per year (1–3 hours each) • Initial face-to-face meeting of all members, 2 days • Monthly update to Steering Team

Use of Opportunity Matrix to Prioritize Gap Closure Activities

Using the principles defined in the opportunity matrix, which compares value creation and complexity of each activity to be undertaken by the Network in closing gaps identified, the opportunity matrix helps in prioritizing MOC Network work as shown in Figure 12.1. By leveraging experience and knowledge in the field, the Core Team allocates scores for value created and complexity for activities associated with closing each gap identified.

Once the gap closure strategy has been prioritized, a more detailed work plan can be developed to assist the Network in stewarding and achieving its goals on schedule and within budgets. Table 12.2 provides a sample work plan for closing MOC gaps identified by the Core Team. A major advantage of the work plan is that the work activities are identified and assigned to the

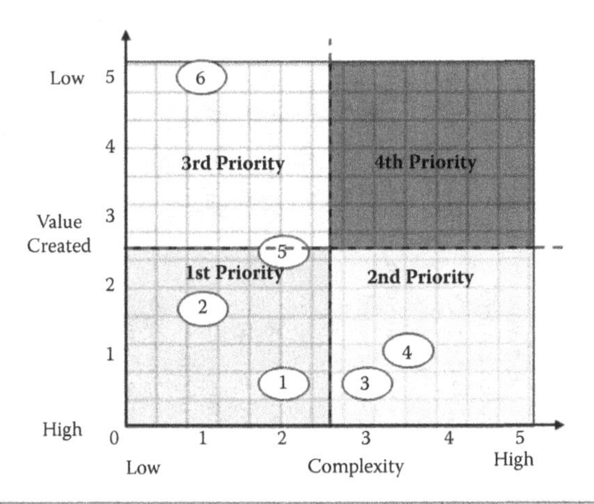

Priority	Opportunity	Complexity 1=Low 5=High	Value Created 1=High 5=Low
1	Build collaboration site for Network/stakeholder	2.0	0.5
2	Resolve MOC-E/MOC-NE definition—key interpretation of standard	1.0	1.5
3	Leverage metrics and audit results from sites into the Network planning	3.0	0.5
4	Select best practice in MOC processes among all business units for corporate sharing	3.5	1.0
5	Select MOC tool for corporate-wide use and execute accordingly	2.0	2.5
6	Update the MOC standard based on improvement opportunities identified	1.0	5.0

FIGURE 12.1
Prioritized gap closure activities for the MOC Network.

TABLE 12.2

Sample MOC Work Plan for Closing Gaps Identified and Prioritized as per the Opportunity Matrix

Priority	Objective	#	Tasks	Deliverables	Owner	Start Date	Finish Date
1	Build collaboration site for Network and stakeholders	1	Define the documents and information that will be shared and communicated to the extended Networks, and what will reside on the collaboration site		Insert Names	Q1	Q1
		2	Establish and populate the MOC collaboration site	Functional collaboration site		Q1	Q1
2	Resolve MOC–engineered/MOC–nonengineered definition, key interpretation of standard	1	Compare business unit variations in definitions of MOC–engineered/nonengineered	Develop a common interpretation and communicate to all stakeholders		Q2	Q4
		2	Review and update the MOC guidance document as required	Updated MOC guidance document		Q3	Q 4
		3	Develop implementation plan to enable the update of related documents/training packages and communicate the changes	Implementation and communication plan		Q3	
3	Leverage metrics and audit results from sites into the Network planning	1	Share findings from Firebag audit related to MOC and PSSR (benchmarking site performance)	Presentation to team		Q2 2012	Q3
		2	Determine if business unit self-assessment reports provide positive value to the MOC improvement process	Assessments of value of business unit self-assessments of their MOC processes		Q3 2012	

(continued)

TABLE 12.2 (CONTINUED)

Sample MOC Work Plan for Closing Gaps Identified and Prioritized as per the Opportunity Matrix

Priority	Objective	#	Tasks	Deliverables	Owner	Start Date	Finish Date
		3	Identify key strengths and opportunities from business units from prior audits and share with the Network to aid in Network planning and gaps identification	List of strengths and weaknesses across the entire organization in MOC process execution		Q4 2012	
		4	Identify and standardize meaningful MOC metrics and KPIs across organization; ensure the MOC tools are capable of reporting these metrics	Standardized MOC metrics used by all business units of the organization		Q2 June	Q3
		5	Participate in business units audits relating to MOC	Network members to participate in a minimum of two PSM MOC audits per year		Q1 2012	Q4
4	Select best practice in MOC training processes among all business units for corporate sharing	1	Review available training materials for best practice potential	Selected MOC process and execution training materials		Q1 2012	
		2	Verify training materials address requirements for PSM and management system	Completed training materials that meet the needs of both PSM and the organizational management system		Q3	
		3	Make accessible to all business units training materials for MOC	Best practice accessible to all business units and being used		Q3	

(continued)

TABLE 12.2 (CONTINUED)

Sample MOC Work Plan for Closing Gaps Identified and Prioritized as per the Opportunity Matrix

Priority	Objective	#	Tasks	Deliverables	Owner	Start Date	Finish Date
5	Select MOC tool for corporate-wide use and execute accordingly	1	Consider an MOC–nonengineered catchall category	Decision on category for global design		Q2 May	5/25/2012
		2	Define *simple* and *complex* and how it relates to engineered and nonengineered change (section 3)	Definition and guidance for global design		Q2 May	Q3
		3	Provide clarity on the pick list options for risk ranking. Determine if risk ranking required for all MOCs. If not required how do we define or when do we use the other options?	Decision and guidance for global design or for inclusion in training and guidance document		Q2-Q3	Q3
		4	Define a minimum set of planning questions for each category of change that site could use/refer to	Set of questions for provider, full-deployment as part of overall reliability improvements implementation package		Q3	Q3
6	Update the MOC standard based on improvement opportunities identified	1	Identify proposed changes and supporting logic	Changes required to standard		Q2	
		2	Evaluate validity and justification of proposed changes	Defensible process for proposed changes		Q2	
		3	Update standard and complete approval process	Updated MOC Standard		Q2	
		4	Yearly review of MOC standards and guidelines	Improved MOC-related documents that reflect current business processes and enable an effective MOC process		Yearly	

team members with due dates. This will ensure that the deliverables listed in the charter will be delivered on time.

Communication Plan and Messaging

Ongoing communication with key stakeholder groups help in keeping everyone informed on the progress of the Network and on upcoming activities. A simple communication brief, as shown in Figure 12.2, helps in achieving this

(Insert company logo here)
Management of Change (MoC) Network Update

At a Glance
Core Team Members:
- List members here

Contact Information
List Team Lead contact details here

Collaboration Room
Insert link to MOC Network collaboration site here

The purpose of this communication brief is to provide high level information regarding the MOC Network activities. This Network team has been at work since October 2011. Team members meet monthly via teleconference and a face-to-face meeting took place during July 2012.
Goal: Suncor Excellence MOC/PSSER Network (SEN) is responsible to maintain the related standards, guidance documents, audit protocols, and share best practices.

Highlights:
MOC Standard, Audit Protocol, and related tools and templates can found on the MOC collaboration site. Access these tools via attached link (insert link here).

Recent Activities:
The network has developed an MOC procedure for consideration by all sites. This was requested by many business areas in an effort to provide:

- A workable solution to sites currently implementing or improving the MOC process.
- A more standardized approach that aligns with MOC management systems requirements.

Updates to the MOC PSM Standards
While the standards are not yet available to be updated, a workshop was held to draft changes to the existing MOC Standards. This is an effort to be ready for the standard revision process. This workshop reaffirmed strengths in the existing standard while identifying potential improvements. This work is considered in draft and has no impact on business areas until the standard revision process begins. However, in the meantime, if any sites wish to suggest changes to the standard, they are encouraged to contact any member of the MOC Network.

Collaboration and Assistance
Now that the network has been activated, it is available to provide support to all BUs.
Support provided includes:

- Interpretations of the requirements of the Standards.
- Sharing best practices from others, across all Business Units and Areas.
- Functions as a facilitator by connecting sites with strengths in MOC with sites that are having difficulties in executing the MOC Standard.

FIGURE 12.2
Sample MOC communication brief for sharing information with stakeholders.

goal. When major changes are proposed for execution across the organization, the MOC process itself must be applied in such cases supported with an appropriate communication plan highlighting the following:

- Types of changes that are being proposed
- Stakeholder groups that are affected by the proposed changes
- Impact of the changes on stakeholders
- Steps taken to mitigate the impact of the proposed changes
- Value created to the organization from the proposed changes

Proper communication media choice is required to ensure changes are properly communicated to stakeholders. It is important to note that when change is proposed, the corporate organization is often accused of tossing the change across the fence for the business unit to execute without real consideration of the business impact to the business unit. Networks must ensure adequate attention is applied to remove this misnomer. A communication plan must therefore ensure the right communication medium and approach is applied that removes any perception of tossing change across the fence and to work with the business in the execution of the change.

Stewardship and Communicating Up the Organizational Chain

Stewardship generally requires the Network to provide at least monthly or quarterly progress reports to senior leadership of the organization. Figure 12.3 demonstrates the use of the stewardship template for providing a snapshot view of the MOC Network status as of the date indicated.

The information provided in the stewardship report indicates that the MOC Network is on track to meet its deliverables. Some minor concerns have been identified requiring support from the Steering Team to resolve. This stewardship report provides information as follows:

- Overall status of the Network on its deliverables
- Status of the sub-initiatives as prioritized by the opportunity matrix
- Planned work and focus for the upcoming reporting period
- Areas of support required from the Steering Team
- Some of the successes of the Network for the current reporting period
- Network participation date, which highlights some of the work-related challenges in getting people involved and participating toward finding solutions for the business-related MOC challenges

Network–MoC **Date: September 14, 2012** **Lead:** (Insert Network lead's name here)	
Overall Status	On target
Sub-Initiative Updates(s)	• Collaboration site completed, tested, and activated across all business units. • Community of practice members identified and initial communication sent to members. • No clear agreement on definitions of MOC-E/MOC-NE. Continue to work with SMEs for agreement. • MOC practices and procedures from three of five BUs received and under review by core team for organization-wide application.
Next Period Outlook	• Monitor use of collaboration site and work with community of practice to maximize value of collaboration. • Obtain practices and procedures from remaining business units for evaluation. • Obtain industry MOC practice and procedures for best practice comparison. • Conduct face-to-face core team review of practices and procedures to identify MOC best practices. • Determine 3–5 high value KPIs for MOC stewardship and performance management. • Select and rollout MOC best practices across all BUs.
Support Requirements	
Steering Team	• Need help in securing SME from BU 1. • Support required in getting BU 2 leadership onboard in use of collaboration site within BU.
Issues and/or Challenges	• Difficulties in accessing SMEs given current workloads. • Leveraging technology for communication and meetings is working to an extent. Require a face-to-face meeting to resolve critical issues and decisions.
Network Successes	
• Rollout of collaboration site with good acceptance and use across BUs for sharing MOC information. • Community of practice team members identified and approved. • SMEs identified and responding to the cause. Motivated team of personnel working together. • All members of the core team in place. • Charter finalized and approved by steering team.	
Network Participation	
Meeting Attendance	
% of Network absenteeism	70%
# of meetings cancelled	2
# of Network vacancies	Zero

FIGURE 12.3
MOC progress update to senior leadership.

When used properly, for several Networks, common themes can be identified for action at the senior levels of the organization.

Conclusion

Like every business process, stewardship is an integral part of the process for success. The success of Networks depends upon

- The level of support received
- The commitment of its members
- The value created by Network activities for the business
- How we communicate the activities and successes of the Network
- The stewardship of the Network

In this practical example, we saw the application and use of several simple tools described throughout the various sections of the book.

Conclusion

Why are Networks exceedingly important to us today? Although the formalized Network is a relatively new concept in the workplace, Networks—formal and informal—have existed for decades and have been known to create improvements wherever they have existed. Today, the need for Networks in businesses is overpowering, driven by:

- Intense competition
- Dwindling and tightly controlled access to resources
- An ever-changing business environment
- Increasing stakeholder demands to do more for less
- The ever-growing challenge for continuous improvements in business

Therefore, Networks are fast becoming an integral part of continuous improvements for any business. More important, the quicker organizations are introduced to the concept of formal Networks, the greater the likelihood for business successes.

When Process Safety Management (PSM) and Management Systems Networks are properly activated with the right organizational structure, priorities, and tools, exceptional business performance can be generated. There are fewer major incidents and a general improvement in health, safety, security, and environment (HSSE) performance and safety culture of the organization. Establishing and stewarding Networks is not a difficult task once a proper methodology is applied. Networks must be properly resourced and supported to achieve its chartered deliverables. With several Networks working together, the need for adequate stewardship is essential. More important, nonauthoritative leadership behaviors are essential, since Network members are largely mature, experienced, and motivated workers who seek opportunities to continually improve the specific PSM areas of focus for the collective improvements across the entire organization.

Networks help in determining the right key performance indicators (KPIs) and performance targets that organizations should steward to. Through prioritization and stewardship, Networks are able to actively close gaps in the business that predispose the organization to unacceptable risks. In addition, where available, Networks provide the right framework for generating continuous improvements in business processes and operations that would have otherwise been deemed acceptable by the organization.

The model and processes identified in this book for the successful activation and sustainment of Networks is by no means unique and groundbreaking. Rather, the authors seek to present formalized Networks in businesses as a new concept that has the potential to support businesses in delivering exceptional business performance. The authors hope that the information and knowledge provided in this book will encourage businesses to embrace the concept of Networks to continuously improve performance in each of the PSM and Management System elements stewarded by the organization.

Index

D

E

F

G

H

O

Printed in the United States
by Baker & Taylor Publisher Services